THE UNIVERSALITY OF PHYSICS

A FESTSCHRIFT IN HONOR OF DENG FENG WANG

THE UNIVERSALITY OF PHYSICS

A FESTSCHRIFT IN HONOR OF DENG FENG WANG

Edited by

Ramzi R. Khuri

Baruch College, City University of New York
New York, New York

James T. Liu

University of Michigan
Ann Arbor, Michigan

Feng Chen

Texas Instruments
Dallas, Texas

and

Wenbiao Gan

New York University
New York, New York

Springer-Science+Business Media, LLC

Library of Congress Cataloging-in-Publication Data

The universality of physics: a festschrift in honor of Deng Feng Wang/edited by Ramzi R. Khuri ... [et al.].

p. cm.

Includes bibliographical references.

ISBN 978-1-4613-5144-3 ISBN 978-1-4615-0585-3 (eBook)

DOI 10.1007/978-1-4615-0585-3

1. Physics—Congresses. I. Wang, Deng Feng, 1965–1999 II. Khuri, Ramzi R. III. Deng Feng Wang Memorial Conference (2000: Princeton University)

QC1 .U547 2001

530—dc21

2001038771

The Volume consists of:

Part I. Proceedings of The Deng Feng Wang Memorial Conference, held August 12, 2000, at Princeton University, Princeton, NJ

Part II. Reprints of some of Deng Feng Wang's most important papers

ISBN 978-1-4613-5144-3

Originally published by Kluwer Academic/Plenum Publishers, New York in 2001

Softcover reprint of the hardcover 1st edition 2001

http://www.wkap.nl/

10 9 8 7 6 5 4 3 2 1

A C.I.P. record for this book is available from the Library of Congress

This book is dedicated to the memory of Deng Feng Wang, loving husband, great friend and brilliant collaborator, whose tremendous personal and professional influence endures far beyond his all too brief life.

Preface

Deng Feng Wang was born on February 8, 1965 in Chongqing City, China. In 1981, at the age of sixteen, Deng Feng was admitted to the Electrical Engineering Department at Tsinghua University, where he demonstrated his extraordinary problem-solving skills in physics and mathematics by obtaining the first prize in a school-wide mathematics competition.

As one of the top students in the China-US Physics Exchange Program, Deng-Feng enrolled in the Princeton University Physics Department in 1986. He distinguished himself early by obtaining one of the highest scores in both the preliminary and general examinations, the former only one month after arriving at Princeton. Deng Feng studied theoretical condensed matter physics under Nobel Laureate Philip W. Anderson and obtained his Ph.D. in 1993 with world-renowned physicist Piers Coleman. During this period he spent one year of research at the Institute of Theoretical Physics at the Chinese Academy of Science. In his Ph.D. dissertation, Deng Feng made important contributions to the field of low-dimensional, strongly interacting electronic systems. After completing his dissertation, Deng Feng joined the Institute of Theoretical Physics at the Swiss Federal Polytechnic School at Lausanne as a post-doctoral fellow with Professor Christian Gruber's group. Deng Feng published more than 20 papers in leading journals, such as the Physical Review. In 1996, the Institute presented him with an achievement award for exceptional contributions to theoretical physics.

In 1997, Deng Feng and his wife, Jing, moved to Waterloo, Ontario, where he began a new career in mathematical finance. Within eighteen months, he not only completed all degree requirements, but also published several research papers on mathematical finance and became an expert in this field.

After four months as an intern at Toronto Dominion Bank in 1998 and another four months as a full time employee at TD Securities Inc. (a subsidiary of Toronto Dominion Bank) in 1999, Deng Feng left for New York City in August 1999, with the hopes of obtaining a position as a quantitative analyst on Wall Street. While swimming with friends in the Atlantic Ocean off Island Beach State Park, New Jersey, Deng Feng was caught in a Rip Current and drowned, tragically, at the age of 34. A memorial service was held in honor of Deng Feng on August 24th at the University of Waterloo. The funeral was held on August 28th at Princeton.

Almost one year later, on August 12th 2000, The Deng Feng Wang Memorial Conference was held in Princeton University, and in which Deng Feng's collaborators and friends presented scientific talks in a testimonial to his tremendous influence on their work and careers. The first part of this volume represents contributions from the conference, and the second part consists of reprints of some of Deng Feng's most important papers.

Deng Feng's untimely death tragically cut short his career, and so we will never know what further contributions to science he may have made. However, if we take the true measure of a person to be not in achievements alone, but more so in how his life affects others, then it is clear that Deng Feng Wang's life has had a profound impact, on both the personal and professional lives of his friends and associates. In the hearts and minds of those of us who were fortunate enough to have known him, he lives on as a strong and noble spirit, a generous soul and a wonderful, creative person, who loved life and enriched the lives of us all.

We gratefully acknowledge the help of Harley Pretty, Ray Laue and Kathy Patterson for logistical support in the organization of the conference at Princeton University. Publication of the proceedings is made possible by support from NSF Grant RUI-9900773, DOE Grant DE-FG02-95ER40899 Task G, NIH Grant NS41846-01 and an Ellison Foundation Young Investigator Award.

THE EDITORS

Contents

Preface vii

Part I Proceedings of The Deng Feng Wang Memorial Conference, held August 12, 2000, at Princeton University, Princeton, NJ 1

RVB Redux: A Synergistic Theory of High T_c Cuprates 3
Philip W. Anderson

Doing 1D Physics with Deng: How the Vision Goes On 9
P. Coleman

Kondo-Lattice Models with Infinite Range Hopping 23
Christian Gruber

The One-dimensional t-J Model with Long Range Interaction 35
James T. Liu

Stability and Symmetry Breaking in Metal Nanowires 41
Charles A. Stafford

From Magnetic Flux and Incommensurability to NMR and Oil Wells 53
Denise E. Freed

Geometry, Quantum Field Theory and NMR 67
Scott Axelrod

Black Holes, String Theory and Fundamental Physics 85
Ramzi R. Khuri

Design of a Low Power Passive Sigma Delta Modulator 93
Feng Chen

Labeling the Nervous System with a Ballistic Approach 101
Wen-Biao Gan

Part II Reprints of some of Deng Feng Wang's most important papers 109

Spectrum and Thermodynamics of the One-Dimensional t-J Model with $1/r^2$ Exchange and Hopping 111
D. F. Wang, James T. Liu, P. Coleman

Gutzwiller-Jastrow Wave Functions for the $1/r$ Hubbard Model 121
D. F. Wang, Q. F. Zhong, P. Coleman

Solutions to the Multiple-Component $1/r$ Hubbard model 129
D. F. Wang

Invariants of the $1/r^2$ Supersymmetric t-J Models 135
D. F. Wang and C. Gruber

Exact Results of the One-Dimensional $1/r^2$ t-J Model without Translational Invariance 143
C. Gruber and D. F. Wang

Quantum Duality and Bethe-Ansatz for the Hofstadter Problem on the Hexagonal Lattice 153
C.-A. Piguet, D.F. Wang, C. Gruber

Exactly Solvable Extended Hubbard Model 159
D. F. Wang

Integrabilities of the Long-Range t-J Models with Twisted Boundary Conditions 167
James T. Liu, D. F. Wang

SU($m|n$) Supersymmetric Calogero-Sutherland Model Confined in Harmonic Potential 175
C.-A. Piguet, D. F. Wang and C. Gruber

$1/r^2$ t-J Model in a Magnetic Field 183
James T. Liu, D. F. Wang

Interaction-Induced Enhancement and Oscillations of the Persistent Current 193
C. A. Stafford, D. F. Wang

Generalizing Merton's Approach of Pricing Risky Debt 203
D. F. Wang

Part I

Proceedings of
The Deng Feng Wang Memorial Conference, held August 12, 2000, at Princeton University, Princeton, NJ

RVB REDUX: A SYNERGISTIC THEORY OF HIGH T_C CUPRATES

Philip W. Anderson
Joseph Henry Laboratories of Physics
Princeton University, Princeton, NJ 08544

Abstract From the moment of their discovery the high T_c superconducting cuprates have been a subject of controversy, and even today there is no complete theoretical consensus, let alone agreement as to the correct interpretation of many of the experimental facts. Yet in spite of the appearance of confusion I sense that there is a growing area of agreement among many theorists, and perhaps even more experimentalists, as to the basic outlines of the theory of these puzzling substances, although the terms in which the ideas are couched may appear very different. What I shall try to do here is to express this quasi-consensus in fairly general terms. Then, in a subsequent paper, I will go on along a specific route to a more complete outline of a theory, on which there may be less of a consensus.

When I published my book, in 1996[1], I was convinced of an underlying normal state resembling that which I will discuss here, (the "Luttinger Liquid"), but I thought that the single plane was never superconducting–"dogma V"– and that therefore superconductivity had to be caused entirely by interlayer tunneling. This view was contradicted decisively by the experimental facts about Tl one-layer cuprate, and dogma V has had to be abandoned. Instead, I shall return to something much more like the RVB theory of 1987-89, to views much closer to the mainstream (if any) and regaining a great deal more of the surprisingly satisfactory phenomenology of the RVB. In exchange I have to give up the, apparently, illusory heuristics of the variation of T_c which the ILT provides. This is not to say that ILT does not play a role; in fact the experimental picture requires it to in the bilayer materials which exhibit the resonant neutron mode, but the basic phenomenology can be discussed using a one-layer model.

The present view is the end result of a lot of history, and incorporates ideas borrowed from many people. In order to make some acknowledgement of those ideas, and also to help sort out the confusion we may feel and the misconceptions many of us may have, I will give below a historical introduction. which is a little more complete than is usual for a research paper.

The Universality of Physics: A Festschrift in Honor of Deng Feng Wang.
Edited by Khuri *et al.* Kluwer Academic/Plenum Publishers, 2001.

HISTORICAL INTRODUCTION

Even the basic idea of a spin liquid state of singlet pair bonds is not original with me, having been proposed in various contexts by Landau and by Pauling himself (as a theory of metals), long before I suggested it as a possible state of the CuO_2 planes of the cuprate superconductors on the basis that spins 1/2 in 2D would be particularly susceptible to it. In my first work on the subject, I took advantage of Rice's calculations on the 1D case and a simple transformation to suggest Gutzwiller projection of BCS wavefunctions as suitable model wave functions for describing RVB states in the half-filled band of the Hubbard-Heisenberg model[2]. Kalmeyer and Laughlin[3] also used Gutzwiller projection of mean field states, but of states in a fictitious magnetic field of half a flux quantum per electron.

After a period of much confusion, it became clear that the "flux phase" of Laughlin was identical to the projected BCS state "s+id" of Affleck-Marston [4] and Kotliar[5]. I feel that the latter description had a much clearer physical motivation because the "s" and "d" components of the anomalous BCS self-energy can be motivated from the superexchange interaction, and this physical point plays a big role in the present work.

A crucial aspect of either RVB is that the excitations are necessarily fractionalised electrons, either by virtue of the fractional flux, or in the BCS version neutral fermionised spins, or spinons. Kivelson et al[6] showed that spinons imply a spinless charge excitation which they called the "holon". My group showed that this spin-charge separation (SCS) had many attractive features in understanding experiments on the normal state[7].

At this point a major difficulty, and two important red herrings, entered the picture. The difficulty is defining an RVB state for other than a half-filled lattice: the problem of doping of the Heisenberg model. Our group naively assumed that the BCS pairing would somehow persist and convert itself into superconductivity of the free holes[8], but the mechanism for this process was surprisingly refractory. In a broad sense, that is what actually happens, but we needed a great deal more experimental and theoretical understanding before we could see how. The experimental red flag was the fact that the relevant T_c was of the order of the bandwidth: much too high.

One red herring was the attempt by Laughlin and his collaborators to accomplish the doping in the flux phase by reducing the flux proportionally to the number of electrons[9]. As Shastry had shown[10], the projected flux phase with exactly half a quantum is a real wave function with no loss of time-reversal symmetry, but the new functions did not have this property and the search for time-reversal breaking turned out to be futile. This should have been evident if one looked at the Affleck-Marston equivalence, but was not immediately recognised.

A second red herring was the demonstration, both experimentally by neutron diffraction[11] and theoretically by Halperin et al[12], that the actual ground state of the Heisenberg model is actually not an RVB but Neel ordered. This led much of the community to give up on the RVB and its concomitant non-Fermi liquid ("NFL") aspect in favor of "antiferromagnetic spin fluctuations [13]. This will serve as a catchall term for the attempt, in Laughlin's phrase, to "sum all the diagrams" in some form of more or less conventional theory, but to retain the experimentally obvious link to magnetism. This left those who were convinced that the data required something else with the Sisyphean task of proving a negative, on which we have wasted a lot of breath.

One of the most frustrating aspects of this controversy is that much of it is semantic and notational. In the pre-cuprate times several papers pointed out that an antiferromagnetic spin interaction such as is derived from the superexchange mechanism could be attractive for d-wave singlet superconductivity[14]. This was described in these papers as "spin fluctuation" effect but is nothing of the sort: it is not parallel in any way to the phonon mechanism or to true spin fluctuations such as occur in ^3He, since the interaction is direct and is not mediated, but screened, by bosonic fluctuations. (The original paper on such effects was Anderson and Morel, 1960).

Of course, there are thriving schools which are capable of denying the centrality of the connection to magnetic concepts such as antiferromagnetism and the Hubbard model, schools which have gained the adherence of a number of senior figures but otherwise not advanced the subject over time. If one needs direct experimental proof that this view is untenable, one need look no farther than the recent demonstrations by Zhang[15] and by Chakravarty[16] that the condensation energy and the condensate fraction are quantitatively correlated to changes in the magnetic neutron resonance in at least two compounds.

Actually, a number of groups continued trying to make progress going on from the original RVB ideas. Our group at Princeton eventually focused on the "s" RVB and the optimally doped case where this describes the normal state well, and added to it a superconductivity mechanism dependent on interlayer interaction[17]–unsucccessfully, in the end, as I described above[18]. Much more purist were the group around Lee, including Wen, Nagaosa, and others [19], who continued to try to work with the gauge theory techniques pioneered by Zou and Baskaran[20] for a formal theory of SCS, and over the years made considerable progress with them; I will borrow many of their ideas here. The gauge theory approach was also followed by Wilczek and collaborators[21].

Laughlin's group, like ours, continued to try to understand experimental results in terms of fractionalisation, but mostly based on flux phase ideas. Other groups which did work of importance along these lines were Fukuyama and collaborators[22], the Houston group of Ting and others[23], and T. M. Rice's group in Zurich[24]; even this is not a full list[25]. I should also particularly mention the work of Kotliar and Liu[26], which is very close to the present point of view.

What revitalised the old RVB theories was a couple of experimental facts. The first, oddly, was first publicised by the advocates of AF spin fluctuations as proof positive of their point of view–the demonstration that the superconducting order parameter had d-wave symmetry[27]. But in fact, five years previously Kotliar and Affleck's optimised RVB's had predicted d-wave, and eventually most of us made the connection that d-wave is an inevitable consequence of the repulsive U Hubbard model. The other fact was the growing evidence–which had been accumulating since 1988, but became inescapable when seen in ARPES–of the spin gap or pseudogap in underdoped samples[28]. When this, too, turned out to have d-like symmetry, it seemed too like the Baskaran et al gauge fluctuating RVB order parameter to be a coincidence. (Although, I have to confess, I was one of the slowest to catch on.)

In the meantime, acceptance of the NFL nature of the normal phases sparked a general desire to explain it–in the view of some of us, to "explain it away." This led to what might be called a "pink herring" on the theory side, the search for "quantum critical points," the idea being that near such a critical point large quantum fluctuations cause unusual phenomena and complicated phase diagrams (but not SCS, to my understanding.) There are various critical points in the phase diagram which will turn out to be important, but in my view none of them play this role–and the most publicised, the "SO5" of Zhang, is not there at all.

Gradually, it became clear that there was a T* crossover where the pseudogap appeared, roughly extrapolating toward the mean field Neel temperature of the insulator, and at least three groups proposed very similar phase diagrams based upon the idea that there were two crossing phase transition lines–the "X" diagram. The other line of the X, a temperature proportional to the doping percentage x, was most sensibly justified by Lee and Wen[19] as representing the phase stiffness energy of the superconductor, while T* was the transition into the RVB state. But for some reason the X diagram is identified with Emery and Kivelson[29], who proposed it on quite different grounds, rather than Fukuyama, who seems to have first published it, or Lee and Wen. In this as in many instances in this history, the course of science was strongly influenced by who caught the eye of scientific journalists.

This more or less brings us up to my starting point. The ideas of T* as representing a transition into a d-wave RVB, and of T_c in the underdoped regime as being a separate transition related to phase stiffness, are as you see implicit in the historical development of the subject. What I shall try to achieve in a subsequent paper is a more synthetic and motivated structure.

References

[1] P. W. Anderson, *The Theory of Superconductivity in the High Tc Cuprates*, Princeton University Press, Princeton N.J., 1997.

[2] P. W. Anderson, Science **235**, 1196 (1987).

[3] V. Kalmeyer and R. B. Laughlin, PRL **59**, 2095 (1987).

[4] I. Affleck and J. B. Marston, PR **B37**, 3774 (1988).

[5] G. B. Kotliar, PR **B37**, 3664 (1988).

[6] S. Kivelson, D. Rokshar, and J. Sethna, PR **B35**, 8865 (1987).

[7] P. W. Anderson et al, Physica **C153-155**, 527 (1988); P. W. Anderson, Physica **C185-189**, 11 (1991).

[8] P. W. Anderson, in *Frontiers and Borderlines of Many-Partcle Physics*, J. R. Schrieffer and R Broglia eds., North-Holland, N.Y., 1988.

[9] R. B. Laughlin, PRL **60**, 2677 (1988) and subsequent papers.

[10] P. W. Anderson, B. S. Shastry, and H. Krishnamurthy, PR **B 40**, 8939 (1989).

[11] B. S. Yang et al, J. Phys. Soc. Japan **56**, 2283 (1987); T. Freltoft et al, PR **B36**, 826 (1987).

[12] S. Chakravarty, B. I. Halperin, and D. Nelson, PRL **60**, 5077 (1988).

[13] T. Moriya, Y. Takahashi, and K. Ueda, J. Phys. Soc. Japan **59**, 2905 (1990); A. J. Millis, H. Monien, and D. Pines, PR **B42**, 167 (1990).

[14] K. Miyake, S. Schmidt-Rink , and C. M. Varma, PR **B34**, 6554 (1986); D. J. Scalapino, E. Loh, and J. E. Hirsch, PR **B34**, 8190 (1986).

[15] S. Chakravarty, PR **B61**, 14821 (2000).

[16] S. C. Zhang and E. Demler, Nature **396**, 733 (1998).

[17] J. M. Wheatley, T. C. Hsu, and P. W. Anderson, Nature **333**, 121 (1988); see also ref. [1].

[18] A. A. Tsvetkov et al, Nature **395**, 360 (1988).

[19] X. G. Wen and P. A. Lee, PRL **76**, 503 (1996), cond mat 9702119; P. A. Lee, N. Nagaosa, T. K. Ng, and X-G Wen, PR **B57**, 6003 (1998); P. A Lee and N. Nagaosa, cond mat 9907019.

[20] G. Baskaran, Z. Zou, and P. W. Anderson, Solid State Comm. **63**, 973; G. Baskaran and P. W. Anderson, PR **B37**, 580 (1988).

[21] summarized in F Wilczek, "Fractional Statistics and Anyon Superconductivity", World Scientific (1990).

[22] Y. Suzumura, Y. Hasegawa, and H. Fukuyama, J. Phys. Soc. Japan **57**, 401, 2768 (1988); H. Fukuyama, Physica Scripta **T27**, 63 (1989), and related papers.

[23] Z. Y. Weng, D. W. Sheng, Y-C Chen, and C. S. Ting, PR **B55**, 3894 (1997).

[24] F. C. Zhang, C. Gros, T. M. Rice, and H. Shiba, Supercond. Sci. & Technology **1**, 36, (1988) and papers with Lederer, Poilblanc and others.

[25] Others were A. Zee, S. John and M. Berciu, P. B. Wiegmann, ...

[26] G. Kotliar and J. Liu, PR **B38**, 5142 (1988).

[27] D. A. Wollman, D. J. Van Harlingen, W. C. Lee, D. M. Ginsburg, and A. J. Leggett, PRL **71**, 2134 (1993); D. A. Brawner and H. R. Ott, PR **B50**, 6530 (1994).

[28] J-C Campuzano, H. Ding, M. R. Norman, M. Randeria, in *Physics and Chemistry of Transition-Metal Oxides*, Fukuyama and Nagaosa eds., p.152, Springer-Verlag Berlin,

1998 is an excellent description; this relies on many earlier papers by the same group and by the group of Z-X Shen., e g D. M. King et al, J. Phys. Chem. Sols. **56**, 1865 (1995); H. Ding et al, Nature **382**, 51 (1996).

[29] V. J. Emery and S. Kivelson, Nature **374**, 434 (1995).

ACRONYMS

ILT –interlayer tunneling theory: the idea that superconductivity results from recovery of interlayer kinetic energy which is frustrated in the NFL normal state

RVB—resonating valence bond, the idea that the state may be described as a quantum liquid of valence-bond singlets.

SCS—spin-charge separation, the notion that spin and charge excitation spectra may be distinct in essential ways, as opposed to quasiparticle theory where the fundamental spectrum is the same for spin and charge.

NFL–non-Fermi liquid, referrring to examples like the luttinger liquid (LL) which do not have the exponent values described by fermi liquid (ie quasiparticle) theory.

DOING 1D PHYSICS WITH DENG: HOW THE VISION GOES ON

P. Coleman
Center for Materials Theory
Rutgers University
Piscataway, NJ 08854

Life is so short for an individual ... only the beautiful spirit of human exists for ever.
—Deng Feng Wang

Abstract Deng Feng Wang was my second graduate student. He was a poet-mathematician — a brilliant, emotionally sensitive young man. It is a sad and unusual event to be asked to reflect on one's former student, after an untimely drowning accident. In this short article, I would like to combine some reminiscences about working with Deng, with some reflections on the physics we did together, telling you how the dreams that influenced our work live on today.

1. Introduction

I was first introduced to Deng in the summer of 1988, by Benoit Doucot, then a post-doctoral student with Phil Anderson's research group at Princeton. These were the heady early days that followed the discovery of high temperature superconductivity and the Princeton group was super-charged with creative activity. I remember my first conversation with Deng — he nervously told me that he was having difficulty following the train of thought in Phil's group. He continued to tell me that he would love to work in condensed matter physics, but felt he did not have the temperament to partake in the feverish main-stream of activity on high-temperature superconductivity. Deng nervously asked me whether I would mind him coming up to Rutgers to talk about physics. So it was that the following Fall, Deng began his regular tuesday visits to Piscataway to partake in physics discussions that on and off, would span the next five years. I can't remember when I first told him that I would take him on as a student, but somehow we hit it off from the beginning. Deng's official advisor remained Phil Anderson. I should say that that although they didn't actually work together, Phil's influence was always there — not only through the support he provided to Deng from Princeton, but also because Phil's ideas in physics provided a tremendous source of inspiration for Deng's work in one dimensional physics.

Condensed matter physics at this epoch was dominated by a feverish activity surrounding high-temperature superconductivity. Our community was captivated by these new materials, not just by the high-temperatures at which they became superconducting, but by a general sense that they represented a fundamentally new departure in the physics of metals. For unlike every other known superconductor, cuprate superconductors are ceramic materials: they are poor metals and in the normal state out of which superconductivity develops at liquid nitrogen temperatures, they display properties wholly unlike any other metal.

Amongst their many unusual properties, the linear temperature dependence of the resistivity, the stark two-dimensionality of the conducting copper oxide planes and their close vicinity to an a special insulating state of matter, called a "Mott insulator" stand out as the most dramatic. The temperature dependent resistivity of conventional metals is generally produced by the scattering of electrons off lattice vibrations (phonons) and it normally saturates at temperatures where the scattering mean-free path of the electrons is comparable with the electron wavelength. In the cuprate metals it is the interactions between the electrons that are thought to produce the resistivity, and for still unknown reasons, the resistivity never saturates, but rises with unabated linearity [1], right up to the melting temperature. The electronic structure of the cuprates is made up of two-dimensional layers of copper-oxide, in which spins reside on the copper site. In the "undoped" phase, each copper atom contains one unpaired magnetic moment, forming an electrical "Mott" insulator in which the electrons are localized by the strong mutual interactions that develop when more than one spin occupies a copper site (Fig. 1(a)). High temperature superconductivity develops when these materials are doped with charge carriers. In 1987 this had led Anderson [2] to propose that the high temperature superconductors should not be regarded as conventional metals, but as "doped Mott insulators" in which spin-less charge carriers are added to a quantum-mechanically disordered fluid of spins (Fig. 1(b)). This "hole in a spin liquid" picture had already begun to exert an immensely strong influence on the thinking of the condensed matter community in the late eighties.

It was against this backdrop that Deng's thesis work developed. We were amongst a community of physicists who sensed that a threshold of discovery had been crossed, opening up a new world of discovery. Deng's first year

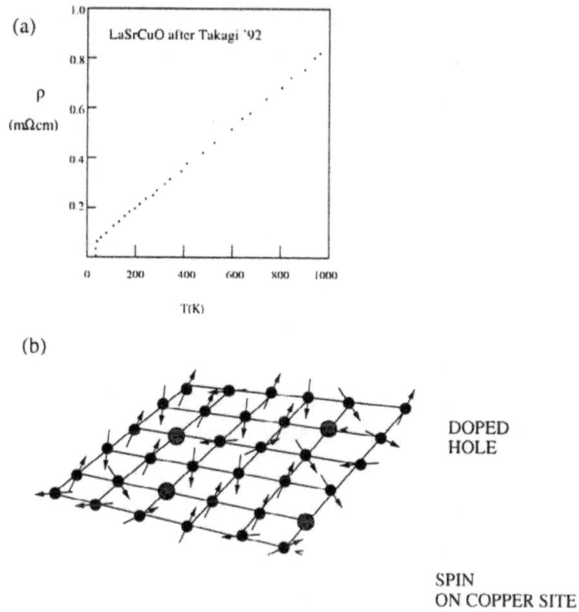

Figure 1. (a) illustrating the linear resistivity in the cuprate metals. (b) According to one picture of the cuprates, high temperature superconductivity is produced when holes are doped into a disordered fluid of spins.

working with me was a learning experience for both of us. From our early conversations, I had already made the mental note to take him on as a student. During this time, we talked a lot about the pervasive belief, held to this day, that the the cuprates are not conventional "Fermi liquids", where the excitations are electron quasiparticles We discussed the possibility that instead, they might be but fluids of pre-formed Cooper pairs [3, 4, 5] which had not yet developed the quantum mechanical phase coherence required for superconductivity. Deng and I came to be driven by the notion that the unusual linear resistance in the normal state of the cuprates could be understood as the result of a diffusion of charge carriers in a fluid of pre-formed pairs. By the Einstein relationship, the electrical conductivity of a material is proportional to a product of the diffusion constant D and the charge susceptibility,

$$\sigma = D\chi_c. \tag{1}$$

We were very much influenced at the time by the notion that pair formation might be closely analogous to the formation of local magnetic moments. Just as the spin susceptibility of a system of magnetic moments is proportional to $1/T$, we reasoned that a system of independent, yet almost localized electron pairs would have a charge susceptibility $\chi_c \propto 1/T$, so that if the charge diffusion constant D were temperature independent, this would lead to

$$\sigma \propto D/T, \qquad \text{or} \qquad \rho = 1/\sigma \propto T, \tag{2}$$

giving a linear resistance. I think many people had had similar thoughts. We tried, without success during the next year to make these ideas into a more tangible theory. Deng worked hard trying to use the Schwinger boson [6] approach to anti-ferromagnetism to develop a theory for charge diffusion in a fluid of pre-formed pairs, but it really did not seem to give the constant diffusion rate we had sought. I should mention as an aside, that at this stage our community was not aware of the "spin gap" [7, 8] that is known to develop in under-doped cuprates at low temperatures, below which a gap, the same size as the superconducting gap is seen to develop. This evidence leads many to believe that pre-formed pairs may form beneath this spin gap, and thus the linear resistivity above the spin gap temperature is very likely to be a property of incoherently scattered, yet unpaired electrons.

By the end of his first year working together, Deng appeared deeply unhappy: after an absence of two years, he missed China and felt desperately lonely. It was clear that he really could not work effectively, and so we came up with the idea that he consider taking a years leave of absence. Deng liked the idea, and thanks to co-operation of Princeton and the help of Yu Lu, at the ICTP in Trieste, we were able to arrange for Deng to go and spend a year in Beijing, supervised by Professor Su at the Academica Sineca. Deng left for China in the early summer of 1990. In the Fall I received a long letter from him that he had written in June, from which I include a few excerpts. He writes:

> *I arrived in Shanghai on the 11th June around midnight, with two other Princeton Chinese graduate students. The flight from New York to Shanghai was 22 hours long, much longer than I can remember....*

Deng's first impressions of China were not entirely rosy — he wrote

> *The first day when I saw the people here, I felt like I wanted to smile to everyone passing in the street. But the city is so unfriendly. The pollution is so bad.... When I first saw the people here, I felt like I could (always) see depression on everyone's face. It looks as if everyone has something unhappy in his life....*

yet he was also philosophical about his return, particularly his reaction to meeting young students at Tsinghua University, where he studied as an undergraduate:

> *The school reminds me of the days of my old dreams, dreaming of making a wonderful country, dreaming of creating a wonderful world, dreaming of helping the people to live a better life here. When I see the young kids passing by in the same schools, where I used to be one of them, I feel like I see the image of my past. Maybe they have the same dreams as I used to have? . . . I always (used to) say to myself "How many roads must a man walk, down before they call him a man? Life is so short for an individual, 80 years or ninety years, only the beautiful spirit of human exists for ever.*

But Deng's arrival in China had also uplifted him, and he reflects later in the letter that

> *I hope that I can make myself a great thinker and can create new spirit in this old land and that the spirit can be such a beautiful thing. But the life is not as beautiful as that spirit at all. I start to think that when one wishes too much, he also suffers too much.*

And finally, at the end of the letter, he writes

> *I am managing to start to work on some physics. I have the chance to go to Princeton to study, the chance which other Chinese young students will never have in their life. I should not waste my time any more.*

These quotes show much about the man Deng was — a deeply thoughtful and sensitive individual, a man with dreams.

I received a number of subsequent letters from Deng that year, generally a mix of physics and philosophy. In one letter, that I regret not having kept, he speculated about whether he should become a Monk, and I became worried that he would not return. Yet Deng did succeed in finding himself during his year of leave and it was a much happier and more self-assured individual that eventually greeted me back upon his return to New Jersey in September 1991. He proudly told me that he had gotten married to Jing, and he felt inspired to work.

2. Working on the Thesis

In the year that had passed, the field of condensed matter physics had moved on. Much more was known about the high temperature superconductors, and this had led to new ideas. An increasingly prominent idea , made by Anderson [9] was that one dimensional physics might provide the key to the normal state physics of the the two dimensional cuprate metals. One dimensional conductors provide one of the few examples of a metal where the charge carriers are not electrons. Unlike a conventional metal, in which charge and spin are carried by electrons, in a Luttinger liquid, the motion of charge and spin are decoupled to form a new kind of metal, called a "Luttinger liquid" [10, 11, 8] Anderson had already suggested that the phenomenon known as "spin charge decoupling" might extend beyond one dimensional metals, to two dimensional conductors [2]. In the early nineties Anderson took one step further and proposed that one might actually be able to regard the Fermi surface of the cuprate metals as a series of one-dimensional conductors dominated by forward scattering — a "Tomographic Luttinger liquid" [9].

Anderson's idea proved a tremendous stimulus to research and led to a renaissance of interest in the physics of one dimension. It was against this back-drop that Deng's thesis took shape. What is the physics of a doped one-dimensional Mott insulator? In 1988, Duncan Haldane [22], and Sriram Shastry [9], had independently discovered a new solvable model for a one dimensional Heisenberg spin-chain. Unlike the spin-chain model solved by Bethe in 1931, the Haldane Shastry spin chain has long-range hopping in which the antiferromagnetic interaction decayed as $1/r^2$. This model was formally written

$$H = \sum_{i>j} J_{ij} \vec{S}_i \cdot \vec{S}_j, \quad J_{ij} = J/d(i-j)^2, \tag{3}$$

where $d(n) = \frac{N}{\pi}\sin(n\pi/N)$ is the "chord distance" between two points n-units apart on circle a circle of circumference length N. In this model, the ground-state and excited-state wave-functions of the spins could be exactly described by a "Jastrow Wavefunction"

$$|\psi\rangle = \exp[i\pi \sum_{\alpha} x_\alpha] \prod_{\alpha<\beta} d^2(x_\alpha - x_\beta) \prod_{\alpha} S^-(x_\alpha)|FM\rangle, \tag{4}$$

where the $\{x_\alpha\}$ represent the positions of the down spins in a ferromagnetic background, represented by $|FM\rangle$. Remarkably, the energies of these states corresponded to a band of free spinons with a quadratic dispersion. From a mathematical standpoint, the simplicity of the model, and the possibility of solving it without resorting to the intricacies of the Bethe Ansatz made this a very attractive discovery.

From the physics standpoint, one could not help noticing the close similarity between the Jastrow wavefunctions for this model, and the famous Laughlin wavefunction [15] for the ground and excited states of the two dimensional quantum Hall fluid. Could a similar wavefunction be found for the doped one dimensional Mott insulator? I suggested this problem to Deng in the fall and he was fascinated by it. Not long thereafter, we learned of the work by Kuromoto and Yokoyama [16], who had introduced a "supersymmetric" t-J model extension of the Haldane Shastry model

$$H = -\sum_{i<j,\sigma} t_{ij} \left[P c^\dagger_{i\sigma} c_{j\sigma} P \right] + \sum_{i<j} J_{ij} (\vec{S}_i \cdot \vec{S}_j - \frac{1}{4} n_i n_j), \tag{5}$$

in which the hopping and spin-interaction strengths are equal, and both decay as the square of the distance between sites.

$$t_{ij} = J_{ij} = t/d^2(i-j), \tag{6}$$

and double-occupancy is projected out by the operators $P = \prod_{i=1,N}(1 - n_{i\uparrow} n_{i\downarrow})$. Kuromoto and Yokoyama had succeeded in showing that the ground-state was a so-called "Gutzwiller" wavefunction, but as yet they had derived a general set of wavefunctions for the excited states. Deng became convinced that this model could be treated exactly, and he set to work on the problem.

I remember this time vividly. Deng made rapid early progress, and working with James Liu, had already checked that Jastrow wavefunctions were eigen-states of finite size systems. However, a general proof was not yet in sight. In

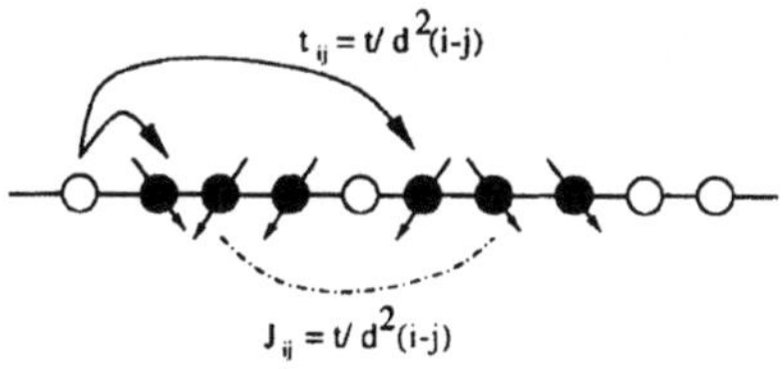

Figure 2. Illustrating the long range hoping of holes and the coupling between spins in the supersymmetric generalization of the Haldane Shastry model.

the Spring of 1992 a set-back occured. We received a paper by Kawakami [17] in which at first sight, he appeared to have completely solved the problem. Norio Kawakami had treated the problem using a method he called the "asymptotic Bethe Ansatz". In this method, the form of the wavefunction is assumed to have the same form as a Bethe Ansatz wavefunction when the spins and holes are far apart. This assumption is self-consistent, and Kawakami used it to obtain the provisional excitation spectrum and thermodynamics of this model.

Deng felt he had been beaten to the solution by Kawakami. I tried hard to console him, but it was initially to no avail: he was in the doldrums. After a week or two in this state, I started to get worried and after a month I really did not know what to do. One day, probably in late April 1992, Deng came into my office in his disconsolate state, and I'm afraid I lost my temper with him. I remember shouting at him the question "Deng: what ARE you going to be doing next year? Your wife is coming, and where are you going to be living? Do you know where?" Deng answered "no", and after a pause I replied "well — you will be living under a tree. How will your wife like to be living under a tree?" I remember telling him to go away and to pull himself together — and as an aside — get a hair cut!

This was one of those actions one takes out of desperation — yet by a miracle, it worked. The next Tuesday, a new person arrived at my office. He was wearing new clothes, smelt sweetly and had had a nice hair cut. More remarkable still, he had with him, a completed derivation of a full set of Gutzwiller Jastrow wavefunctions for the ground and excited states of this model. The Deng Feng Wang wavefunction [18] takes the form

$$\psi_{DFW}(x, y; J_s, J_h) = \exp\left[\frac{2\pi i}{N}\left(J_s \sum_\alpha x_\alpha + J_h \sum_i y_i\right)\right] \Psi_o(x, y), \quad (7)$$

where the $\{x_\alpha\}$ are the co-ordinates of the down-spins in a ferromagnetic background and the $\{y_j\}$ are the co-ordinates of the holes. The quantities J_s and J_h describe the momenta of the down spins and holes respectively, and they take on integral, or half-integral values. Finally, the Jastrow component of the wavefunction takes a form first considered by Laughlin and Kalmeyer

$$\Psi(x) = \prod_{\alpha<\beta} d^2(x_\alpha - x_\beta) \prod_{i<j} d(y_i - y_j) \prod_{\alpha,i} d(x_\alpha - y_i). \tag{8}$$

Deng was able to show that subject to a set of constraints on the two momenta J_h and J_s, his wavefunction was an eigenstate of the long-range supersymmetric Hubbard model. Furthermore, these constraints could be used to re-derive the set of Bethe-Ansatz conditions that had been obtained by Kawakami. Needless to say, I was overjoyed.

In the following year, Deng derived the thermodynamics of the long-range supersymmetric t-J model [18], he also went on to use similar methods to treat a long-range Hubbard model with infinite on-site interaction strength [19], first introduced by Gebhardt and Ruckenstein [20]:

$$H = \sum_{i<j,\sigma} t_{ij} P c^\dagger_{i\sigma} c_{j\sigma} P, \tag{9}$$

where $t_{ij} = it(-1)^{(i-j)}/d(i-j)$. Already by the middle of his last year, it was clear that Deng was happiest doing mathematical physics. Deng's discomfort with the more imprecise, hand-wavy aspects of condensed matter theory had blossomed into a full-blown love of exactly solvable models. Deng went onto work with Christian Gruber, and extended many of the techniques and models that he had come into contact as a Graduate student. I was happy to have been part of this development.

3. Ten Years Later: the Vision Lives On

I would like to end this brief talk telling you how many of the aspirations that inspired Deng's thesis work live on in this new century. Most importantly, we begin to realize that the recent discoveries at the end of the 20th century are represent only the tip of a huge iceberg. Condensed matter physics appears increasingly as a new kind a frontier. We can see that as we move from the elementary metals and elements, to the binary, tertiary and quaternary compounds, the number of stable compounds grows exponentially and as the complexity of the system grows, we believe that new robust kinds of collective behavior emerge. These new phenonmena require new principles, new models and new mathematics for their description, but they are robust against the details of the complexity This is the concept of "emergent behavior" originally introduced by Anderson [21]. The robust regions of phase space where these new phenomena develop have more recently been referred to as "Quantum Protectorates" [22]. From this perspective, the cuprate metals are but a milestone in a frontier where the plain for discovery fans out exponentially from the simplest elements towards the elementary compounds of life.

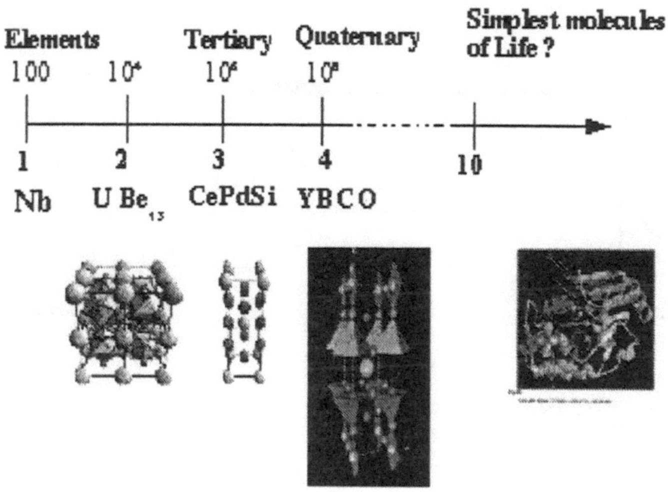

Figure 3. Illustrating the "emergent frontier". As we move from the simplest elements, with one element per formula unit, to more complex binary, tertiary to quaternary compounds, the phase space for new materials grows exponentially. This diagram illustrates some representative materials from each category of compound. (i) UBe_{13} is a heavy fermion superconductor with electron masses in excess of 1000 electron masses. (ii) $CePd_2Si_2$ is a material which can be tuned to a quantum critical point, where it exhibits superconductivity and a quasi-linear resistivity. (iii) Showing $YBa_2Cu_3O_7$, the first superconductor with a transition temperature in excess of the boiling point of liquid nitrogen and (iv) a protein molecule.

A decade after I started working with Deng, the condensed matter community has made huge strides into the understanding of the superconducting state of the high temperature superconductors. Perhaps most dramatically, we now have strong experimental support for presence of nodes in the pair wavefunction, which appears to have a d-wave symmetry $\Delta(\vec{k}) \sim k_x^2 - k_y^2$ [23]. Ironically however, there is still no consensus on the anomalous nature of the normal state. The outstanding problem of the linear resistance is now commonly understood in terms of a "marginal Fermi liquid" [24] transport scattering rate which grows linearly with the temperature

$$\hbar \tau_{tr}^{-1} \approx 2k_B T. \tag{10}$$

There is as yet, no consensus on the origin of this linear scattering rate. Around the time of Deng's thesis, a new twist became apparent. In a conventional metal, the transport relaxation time, which governs the decay of electrical currents also governs the Hall angle $\theta_H = \sigma_{xy}/\sigma_{xx} \sim \omega_c \tau_{tr}$, which in this case would lead

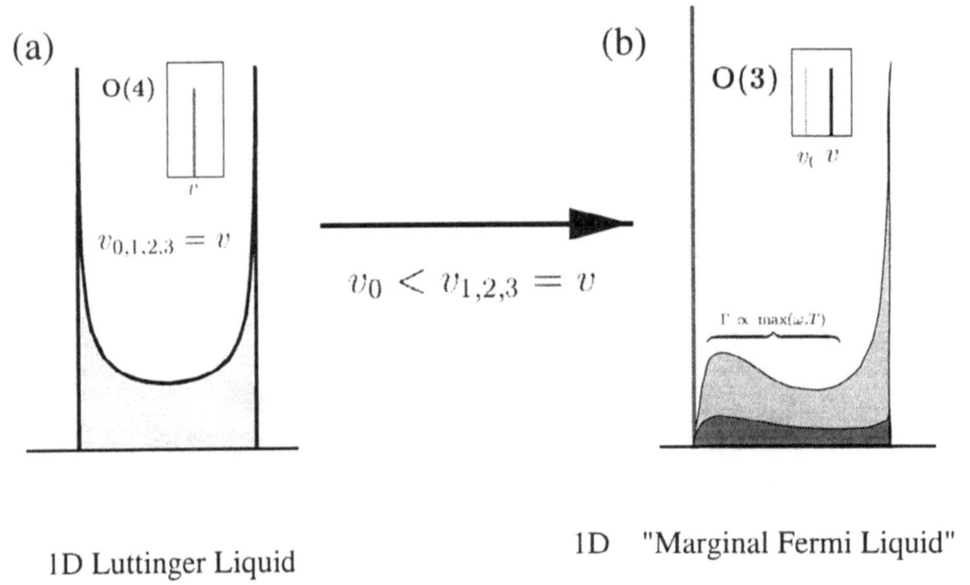

Figure 4. Schematic illustration of the profile of the electron Green's function. (a) The O(4) chiral model, where velocities of all Majorana modes are equal, and the electron Green's function exhibits the standard two-peaked structure of a Luttinger liquid, with a peak at the holon and spinon velocities. (b) The O(3) chiral model has properties in 1D resembling a marginal Fermi liquid. By changing the velocity of one Majorana component, a long-lived quasiparticle splits off from the continuum, which has a width linear in energy.

to a Hall angle $\theta_H^{-1} \propto T$. Experimentally, the inverse Hall angle is found to have a T^2 dependence [25], suggesting Hall currents may decay at a rate which grows quadratically with temperature $\theta_H^{-1} \propto \tau_H^{-1} = aT^2 + b$.

Deng gave me an appreciation of one dimensional physics. Part of the dream with Deng was to try to find new types of electron fluid. Some years later, working with my student Andrew Ho, we did find a new class of one dimensional "electron" fluid [26]. Ho and I asked whether it might be possible to observe a departure from Luttinger liquid behavior in a one-dimensional conductor. The spin-charge decoupling present in one dimension is intimately associated with the kinematics of one dimension. The right (or left) moving electron fields a one-dimensional electron gas can always be divided into two real components,

$$\begin{aligned} \psi_\uparrow(x) &= \frac{1}{\sqrt{2}}[\eta_o(x) + i\eta_1(x)], \\ \psi_\downarrow(x) &= \frac{1}{\sqrt{2}}[\eta_2(x) + i\eta_3(x)]. \end{aligned} \tag{11}$$

In the absence of back-scattering, the physics of the right-moving electrons can then be written in the form

$$H = \int dx \left[-iv \sum_{a=0,3} \eta_a(x)\nabla_x\eta_a(x) + U\eta_o(x)\eta_1(x)\eta_2(x)\eta_3(x) \right], \quad (12)$$

where v is the velocity of the particles. This model has an $O(4)$ symmetry near the fermi energy, corresponding to the conservation of both spin and charge. If we break this symmetry by modifying the velocity of one of the Majorana components is modified, the model reduces its symmetry to an $SO(3)$ symmetry, written

$$H = \int dx \left[-iv_0\eta_0(x)\nabla_x\eta_0(x) - iv \sum_{a=1,3} \eta_a(x)\nabla_x\eta_a(x) \right.$$
$$\left. + U\eta_o(x)\eta_1(x)\eta_2(x)\eta_3(x) \right]. \quad (13)$$

Earlier work on the two relaxation time scales in the cuprate metals had suggested that models with such a split in velocities would display marginal Fermi liquid properties. Ho and I were able to solve this problem exactly, and to show that interestingly enough, the electrons split up into one component with an essentially infinite lifetime, and another with a lifetime that grows linearly with energy. Loosely speaking, we had found a one dimensional electron fluid with two relaxation time scales. Electron fluids of a closely related type may be possible in the edge states of multi-layer quantum Hall systems [27].

In the last five years, he debate surrounding the anomalous normal state of the cuprate metals has widened to encompass a broader class of materials, ranging from one dimensional ladder compounds, to three dimensional materials that lie close to a so-called "quantum critical point" [28]. Research on three dimensional metallic materials has discovered that pressure, or chemical pressure can be used to tune when magnetic metals, or more specifically, antiferromagnetic metals to a "quantum critical point" where the magnetic transition temperature goes to zero. At such quantum critical points, the quantum fluctuations in the magnetization long correlations in both space, and in time. These fluctuations cause the electrons to become far more strongly interacting. Experimentally the metal develops highly anomalous normal state properties — such as a quasi-linear resistivity and a slowly diverging specific heat coefficient. In the case of quantum-critical $YbRh_2Si_2$, a linear resistance is observed over three decades of temperature [29]. It is widely believed that at such quantum critical points, the Fermi liquid behavior of the metallic state breaks down — either through the strong scattering of electrons of critical magnetic fluctuations — or perhaps — still more dramatically, through the complete break-down of the electron quasiparticle.

One of the most dramatic features of this quantum critical behavior, is that it appears that the only energy scale governing the quantum critical fluctuations, is temperature itself. Thus in the quantum critical material $CeCu_{6-x}Au_x$ ($x = 0.1$), [30], the inelastic neutron scattering spectrum at the critical Bragg vector has a scale invariant form

$$\chi''_Q(E,T) = \frac{1}{T^\alpha} F(E/k_B T), \tag{14}$$

where $\alpha \approx 0.8$. This kind of evidence suggests that the underlying physics behind this quantum criticality may involve a hitherto undiscovered critical Lagrangian that lies below its upper-critical dimension, even when the physics is three-dimensional [28]. This is currently an area of furtive interest that affects both the two dimensional cuprate materials, and the almost antiferromagnetic, three dimensional heavy fermion compounds.

I would like to close with a few remarks about Deng. I continued to keep in good contact with Deng and his wife Jing, after he left New Jersey for Switzerland. Towards the end of a highly successful post-doctoral career in Lausanne, Deng made the decision that he would like to apply his mathematical skills to finance, and he moved to Canada to retrain in his new chosen field. Ironically, on the day he died in a swimming accident off the coast of New Jersey, I gather he had been tentatively offered a position at a New York Bank.

The kind of brilliant yet artistic spirit that drove Deng, and inspired him to come from a very humble background, to try his skills in math and physics lives on today. I hope his example will inspire other young people across our globe to try their hand at the frontiers of condensed matter physics research.

Acknowledgments

This work was supported by the National Science Foundation under grant DMR 9983156. I would like to thank Ramzi Khuri for his great patience in waiting for the final version of this manuscript.

References

[1] E.g. M. Gurvitch and A. T. Fiory, Phys. Rev. Lett. **59**, 1337 (1987); L. Forro *et al., ibid.* **65**, 1941 (1990).

[2] P. W. Anderson, Science, **235** 1196, (1987).

[3] S. Schmitt Rink, P. Nozieres, J. L. Temp Physics **59**, 195 (1985).

[4] V. J. Emery and S. Kivelson, Phys. Rev. Lett **74**, 3253 (1995).

[5] V. B. Geshgenbein, L. B. Ioffe and A. I. Larkin, Phys. Rev. B **55**, 3173 (1997).

[6] D.P. Arovas and A. Auerbach, *Phys. Rev.* **B38**, 316 (1988).

[7] J. W. Loram, K. A. Mirza, J. M. Wade, J. R. Cooper and W. Y. Liang, Physica C **235–240**, 134 (1994).

[8] R. E. Walstedt, R. F. Bell, D. B. Mitzi, Phys. Rev. B **44** 7760 (1991).

[9] P. W. Anderson, Phys. Rev. Lett **64**, 1839 (1990).

[10] J. M. Luttinger, J. Math Phys **4**, 1154 (1963).

[11] M. Flicker and E. H. Lieb, Phys Rev 161, 179 (1967).

[12] F. D. M. Haldane, J. Phys C **14**2585 (1981).

[13] F. D. M. Haldane, Phys. Rev. Lett. **60**, 635 (1988).

[14] B. S. Shastry, Phys. Rev. Lett. **60**, 639 (1988).

[15] R. B. Laughlin, Phys. Rev. Lett. **50**,1395 (1983).

[16] Y. Kuromoto and H. Yokoyama, Phys. Rev. Lett. **67**, 1338 (1991).

[17] N. Kawakami, Phys. Rev. B 45, 7525–7528 (1992).

[18] D. F. Wang, J. T. Liu and P. Coleman, Phys. Rev. **B46** , 6639 (1992).

[19] D. F. Wang, Q. F. Zhong and P. Coleman, Phys Rev. **B 48**, 8476–8479 (1993)

[20] F. Gebhard and A .E. Ruckenstein, Phys. Rev. Lett. **68**, 244 (1992).

[21] P. W. Anderson, "More is Different", Science **177**, 393 (1972).

[22] R. B. Laughlin and D. Pines, Proc Natl Acad. Sci, **97**, 28 (2000); **97**P. W. Anderson, Science **288**, 480, (2000); D. Pines, Physica C, **341**, 59 Nov. 2000.

[23] D. A. Wllmann, D. J. Van Harlingen, W. C. Lee, D. M. Ginsberg and A. J. Leggett, Phys. Rev. Lett **71**, 2134 (1993); C. C. Tsuei and J. R. Kirtley, Rev. Mod. Phys. **72**, 969 (2000).

[24] C. M. Varma, P. B. Littlewood, S. Schmitt-Rink, E. Abrahams and A. E. Ruckenstein, Phys. Rev. Lett. **63**, 1996, (1989)..

[25] T. R. Chien, Z. Z. Wang and N. P. Ong, Phys. Rev. Lett. **67**, 2088 (1991).

[26] A. F. Ho and P. Coleman, Physical Review Letters, **83**, 1383 (1999).

[27] J. D. Naud, L. Pryadko, S. L. Sondhi, Nucl.Phys. **B594** 713, (2000).

[28] For a review of these issues, see P. Coleman, C. Pepin, Q. Si and R. Ramazashvili, cond-mat 0105006.

[29] O. Trovarelli, C.Geibel, S. Mederle, C. Langhammer, F.M. Grosche, P. Gegenwart, M. Lang, G. Sparn and F. Steglich, Phys. Rev. Lett. **85** 626 (2000).

[30] A. Schroeder, G. Aeppli, R. Coldea, M. Adams, O. Stockert, H. von Lohneyson, E. Bucher, R. Ramazashvili and P. Coleman, Nature, 407, 351–355 (2000).

KONDO-LATTICE MODELS WITH INFINITE RANGE HOPPING

Christian Gruber
Institut de Physique Théorique
Ecole Polytechnique Fédéderale de Lausanne, PH B Ecublens
CH-1015, Lausanne, Switzerland

Abstract In this talk in honor of Deng Feng Wang I will present a work we did together on the multichannel Kondo-lattice model with constant amplitude for hopping between any two sites of the lattice of the conducting electrons. For this model we obtained the exact thermodynamic properties at zero temperature and the metal-insulator transition. In the limit of infinite coupling constant between conduction electrons and impurities we obtained the ground state energy, which is highly degenerate, and the wave functions.

1. Introduction: Scientific work with Deng Feng at EPFL

During the 4 years Deng Feng worked in Lausanne (1993-1997), his main interest was centered on strongly correlated electronic systems. Due to the strong correlations and low dimensionalities, these systems have to be investigated with non-perturbative methods. Using rigorous approach he investigated the ground state properties, the thermodynamics, and the spin and charge dynamics of various systems in 1 and 2 dimensions.

1. $t-J$ Models with $1/r^2$ interactions [1]. He introduced a new $t-J$ model of Haldane-Shastry type defined on a non-translational invariant lattice. Indeed some previous results had been obtained using explicitly the invariance under translation and he wanted to understand what remains when this symmetry is absent. He obtained the ground state wave function in the form of Jastrow product, the full energy spectrum, and proved the integrability of his model.

2. Electron in a magnetic field [2]. In this area he studied the Bethe ansatz for electron hopping on a 2-dimensional hexagonal lattice. He derived first a duality relation between the hexagonal and the triangular lattice from which he obtained the energy spectrum. He also investigated interacting anyons systems and their quantum groups.

3. Superconductivity [3]. In this domain he was able to show that a lattice model with off-diagonal long range order cannot support a magnetic field. This indicates the existence of a Meissner effect in these systems.

4. Magnetic impurity systems [4]. Several magnetic models were investigated. He was especially interested in the competition between the kinetic energy of the electrons, and their interaction with impurities, which accounts for the metal-insulator transition in heavy fermion systems.

I have chosen to speak on this last domain of Deng-Feng's research because it illustrates very well his interest and some of the methods he used while working in Lausanne.

This article gives me moreover the opportunity to stress that Deng-Feng has always been the motor of our collaboration, coming with new ideas, or suggesting new models to investigate. As a matter of fact, for me the incursion in the field of integrable models was entirely due to the collaboration, and friendship, we developed during his stay in our institute and everything I will mention in the following is entirely due to him. I shall remember Deng-Feng as a real friend who just loved doing physics. Jing, his wife, would say he loved it too much. He was profoundly honest and would never accept any compromise either in physics or in life. For example, he was always very careful to give full credit to those people from which he borrowed some ideas. Of course he expected the same in return and would get very angry if his work was not properly recognized. On another occasion he almost caused a riot at the railroad station in Beijing because he found that the taxi driver wanted to charge us too much. This trip through China with Deng-Feng and his wife gave me the opportunity to appreciate even better his deep personality and culture, as well as the profound attachment he had with his home country.

2. The Multichannel Kondo-Lattice Model

The system under consideration consists of itinerant spin $1/2$ electrons on a lattice, interacting with magnetic impurities through a spin-exchange term. The electrons can be in one or several conduction bands (or channels); the impurities are spin-$1/2$ fermions. It is moreover assumed that there is exactly one impurity at each site of the lattice.

The Kondo-Lattice model is defined as follow. We consider a lattice $\Lambda = \{1, 2, \ldots, L\}$ of L sites; at each site $i \in \Lambda$, we have creation and annihilation orators $(c^+_{i\alpha m}, c_{i\alpha m})$ for electrons with spin $\alpha \in \{\uparrow, \downarrow\}$ in the channel $m \in \{1, 2, \ldots, M\}$, and similarly $(f^+_{i\alpha}, f_{i\alpha})$ for impurities. The Hilbert space $\mathcal{H}$ is the tensor product of Hilbert spaces for electrons and impurities.

The Hamiltonian consists of the kinetic energy of the electrons, and the on-site spin-spin interaction between electrons and impurities, i.e.

$$H = H_0 + JV \tag{1}$$

with

$$H_0 = \sum_{\substack{i \neq j \\ \alpha, m}} t_{ij} c^+_{i\alpha m} c_{i\alpha m} \tag{2}$$

$$V = \sum_i \vec{S}_i^{(e)} \cdot \vec{S}_i^{(f)} \tag{3}$$

$$\vec{S}_i^{(e)} = \sum_{\alpha\beta m} \frac{1}{2} c^+_{i\alpha m} \vec{\sigma}_{\alpha\beta}\, c_{i\beta m} \tag{4}$$

$$\vec{S}_i^{(f)} = \sum_{\alpha\beta} \frac{1}{2} f^+_{i\alpha} \vec{\sigma}_{\alpha\beta}\, f_{i\beta} \tag{5}$$

where $\vec{\sigma}_{\alpha\beta}$ are the Pauli matrices and J the coupling constant. The following *two conditions* are imposed on the model:
1. There is exactly one impurity at each lattice site. The Hilbert space is thus the subspace $\overline{\mathcal{H}} = \mathcal{H}^{(e)} \otimes \overline{\mathcal{H}}^{(f)}$ of $\mathcal{H}$.
2. The hopping t_{ij} is taken to be uniform, i.e. $t_{ij} = -t$ for all $i \neq j$. We thus have

$$H_0 = \sum_{k\alpha m} \epsilon(k) c^+_{k\alpha m} c_{k\alpha m}, \tag{6}$$

where

$$\epsilon(k) = -tL\delta_{k,0} + t \tag{7}$$

and

$$c^+_{k\alpha m} = \frac{1}{\sqrt{L}} \sum_{j=1}^{L} e^{ikj} c^+_{j\alpha m}. \tag{8}$$

Because of the special form of H_0 the dimensionality of the lattice is irrelevant and we have essentially a one dimensional model.

The problems to be discussed in the following concern the zero temperature properties, in particular:

a) the thermodynamic properties in the limit $L \to \infty$;

b) the ground state energy and wave function for L finite.

In the next sections I shall concentrate my attention on the above model, but Deng-Feng investigated also other models with long range hopping such as the chiral Kondo model with $\epsilon(k) = -tk$, i.e.

$$t_{r,s} = -it(-1)^{r-s} \left[\frac{L}{\pi} sin(\pi \frac{r-s}{L}) \right]^{-1}.$$

3. Thermodynamic Properties

As usual one starts with the partition function

$$Z = \mathrm{Tr}_{\bar{\mathcal{H}}} e^{-\beta[H - \sum_m \mu_m N_m]} \tag{9}$$

where μ_m is the chemical potential and N_m the number operator for electrons in the m-th channel. From Z one obtains the pressure

$$p(\beta, \{\mu_m\}; t, J) = \lim_{L\to\infty} \frac{1}{L}\frac{1}{\beta} \ln Z \tag{10}$$

from which one derives the electrons density in channel m, i.e. $n_m(\beta, \{\mu_m\})$, and the energy density $e(\beta, \{\mu_m\})$. Taking the limit $\beta \to \infty$ one obtains the zero temperature values $n_m(T = 0, \{\mu_m\})$ and $e(T = 0, \{\mu_m\})$. To compute the pressure, we can prove the following result, on which I shall come back at the end of this section.

Lemma:

$$p(\beta, \{\mu_m\}; t, J) = p_0(t) + p(\beta, \{\mu_m\}; t = 0, J) \tag{11}$$

with

$$\begin{aligned} p_0(t) &= 2Mt\theta(t), \\ \theta(t) &= \begin{cases} 1 & \text{for} \quad t \geq 0 \\ 0 & \text{for} \quad t < 0 \end{cases} \\ \tilde{\mu}_m &= \mu_m - t. \end{aligned}$$

In the grand canonical formalism, given the chemical potential $\{\mu_m\}$, we have from Eqn. (11)

$$n_m = \frac{\partial}{\partial \mu_m} p = \frac{\partial}{\partial \tilde{\mu}_m} p(\beta, \{\tilde{\mu}\}; t = 0, J) \tag{12}$$

$$\begin{aligned} e_m &= -\frac{\partial}{\partial \beta}(\beta p) + \sum_m \mu_m n_m = \\ &= -2Mt\theta(t) + t\sum_m n_m + e_0 \end{aligned} \tag{13}$$

$$e_0 = J\langle \vec{S}^{(e)} \cdot \vec{S}^{(f)} \rangle_{(L=1)}$$

To study the thermodynamic limit L going to infinity, we thus remark that it is possible to start from a finite system with the equivalent one-body Hamiltonian

$$H^{\mathrm{eq}} = -2LMt\theta(t) + t\sum_m N_m + H^{int} \tag{14}$$

$$H^{\mathrm{int}} = J\sum_i \vec{S}_i^{(e)} \cdot \vec{S}_i^{(f)}, \tag{15}$$

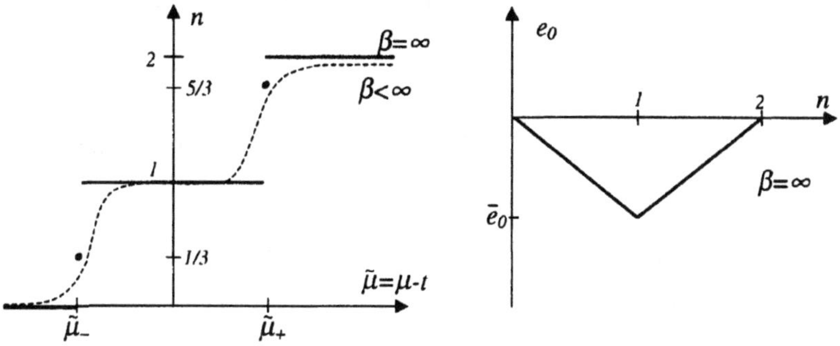

Figure 1. Electrons and energy densities in the grand canonical formalism.

and to use the canonical formalism with N_m, or $N = \sum_m N_m$, fixed. Up to a constant the energy at $T = 0$ is then simply the ground state energy $E_{\rm GS}^{\rm int}$ of the spin exchange interaction $H^{\rm int}$.

We have the following results.

I) *System with one channel* ($M = 1$)

Using the grand canonical formalism, Eqn. (12) and (13) yield the electron and energy densities represented on Fig. 1 with

$$\begin{aligned}\tilde{\mu}_\pm &= \pm\frac{3}{4}J, \quad \bar{e}_0 = -\frac{3}{4}J, \text{ for } J > 0 \text{ (antiferromagnetic)}\\ \tilde{\mu}_\pm &= \pm\frac{1}{4}|J|, \quad \bar{e}_0 = -\frac{1}{4}|J|, \text{ for } J < 0 \text{ (ferromagnetic)}\end{aligned}$$

To understand this result we can use the canonical formalism and the previous remark. We thus need to compute the eigenvalues of the operator $J\, \vec{S}_i^{(e)} \cdot \vec{S}_i^{(f)}$. In this case ($M = 1$) the Hilbert space at each site i has dimension 8; the four states with zero or two electrons on this site have energy zero; as for the four states with exactly one electron at site i we have a single state (of the electron-impurity system) with energy $-\frac{3}{4}J$ and a triplet state with energy $\frac{1}{4}J$. We thus obtain the ground state energy of the equivalent Hamiltonian (14) for a given N:

a) for the *antiferromagnetic* case $J > 0$

$$E_{\rm GS}^{\rm int} = \begin{cases} -\frac{3}{4}JN & \text{if } N \le L, \quad N \text{ sites with } 1e^- \text{ (singlet)} \\ -\frac{3}{4}J(2L-N) & \text{if } N \ge L, \quad (2L-N) \text{ sites with } 1e^- \text{ (singlet)} \\ & \qquad \text{and } (N-L) \text{ sites with } 2e^- \end{cases}$$

b) for the *ferromagnetic* case $J < 0$

$$E_{\rm GS}^{\rm int} = \begin{cases} -\frac{1}{4}|J|N & \text{if N} \le \text{L}, \quad N \text{ sites with } 1e^- \text{ (triplet)} \\ -\frac{1}{4}|J|(2L-N) & \text{if } N \ge L, \quad (2L-N) \text{ sites with } 1e^- \text{ (triplet)} \\ & \qquad \text{and } (N-L) \text{ sites with } 2e^- \end{cases}$$

Using Mattis criteria for the nature, "metallic" or "insulator", of the ground state, i.e.

$$E_{GS}(N+1) + E_{GS}(N-1) - 2E_{GS}(N) = \begin{cases} 0 & \text{for "metallic"} \\ > 0 & \text{for "insulator"} \end{cases}$$

we arrive at the conclusion that:
for any value $J \neq 0$, the system becomes an insulator at half-filling ($N = L$), while it is metallic for $N \neq L$.
(On the other hand the system is metallic for $J = 0$)
II) *System with 2 channels (M=2)*

In this case the Hilbert space at each site has dimension 32 and the eigenvalues of the on-site spin-exchange term

$$J \sum_{i=1}^{2} \vec{S}_{im}^{(e)} \cdot \vec{S}_{j}^{(f)}$$

are the following

- for 0 or 2 electrons in each band (8 states): $E = 0$
- for 1 electron on one band, and 0 or 2 in the other (16 states):

$$E = -\frac{3}{4}J \qquad \text{singlet}$$

$$E = \frac{1}{4}J \qquad \text{triplet}$$

- for exactly 1 electron in each band (8 states):

$$E = 0 \qquad \text{doublet}$$

$$E = -J \qquad \text{doublet}$$

$$E = \frac{1}{2}J \qquad \text{quadruplet}$$

From the above spectrum, we obtain for fixed N_1, N_2 (Number of electrons in channel 1 and 2) the ground states represented on Fig. 2:
For $J > 0$ (antiferromagnetic), the ground state is "metallic" everywhere except at the point ($N_1 = 1$, $N_2 = 1$), and on the four diagonals, where it is "insulator";
For $J < 0$ (ferromagnetic), the ground state is "metallic" everywhere except for the one point ($N_1 = 1$, $N_2 = 1$), where it is "insulator".

For example, for $J > 0$ then in $\mathcal{D}_1$ the ground states have zero or one electron at each site in any of the two bands (singlet) and $E_{GS}^{int} = -\frac{3}{4}JN$, $N = N_1 + N_2$; on the other hand in $\mathcal{D}_2$, the ground states have $(2L - N)$ sites with one electron,

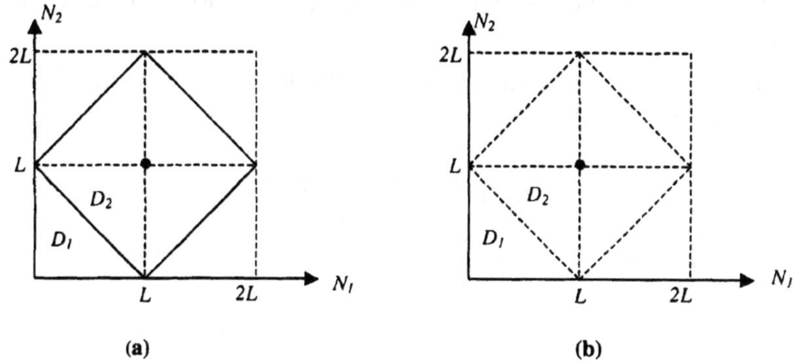

Figure 2. Ground states for given N_1, N_2; the fat lines and the point (L, L) are the points where the system is an insulator.

and $(N - L)$ sites with two electrons, one in each band, and total spin of $e^- = 1$; $E_{\rm GS}^{\rm int} = -\frac{1}{4}JN$; therefore $\frac{\rm d}{{\rm d}N}E_{\rm GS}^{\rm int}$ is discontinuous for $N_1 + N_2 = L$.

However for $J < 0$, in $\mathcal{D}_1$ and $\mathcal{D}_2$ the ground states have one electron in each band; the total spin of electrons plus impurity is $3/2$ (quadruplet): $E_1^{\rm int} = E_2^{\rm int} = -\frac{1}{2}|J|N$ and thus no discontinuity of $\frac{\rm d}{{\rm d}N}E_{\rm GS}^{\rm int}$ for $N_1 + N_2 = L$.

To conclude this discussion on the ground state properties we want to indicate the *main steps for the proof of the lemma*:

1. Replace the trace over the restricted Hilbert space $\overline{H}$ (one impurity at each site) by a trace over the full Hilbert space H. To do so one introduces the impurity isospin

$$\vec{\tau}_i^{(f)} = \sum_{\alpha\beta} \frac{1}{2} \tilde{f}_{i\alpha}^{+} \, \vec{\sigma}_{\alpha\beta} \, \tilde{f}_{i\beta}, \quad \tilde{f}_{i\uparrow} = f_{i\uparrow}, \quad \tilde{f}_{i\downarrow} = f_{i\downarrow}^{+}$$

which leads to

$$Z = \frac{1}{2^L} {\rm Tr}_{\mathcal{H}} e^{-\beta[H' - \sum \mu_m N_m]},$$

$$H' = H_0 + JV', \qquad V' = \sum_j \vec{S}_j^{(e)} \cdot (\vec{S}_j^{(f)} + \vec{\tau}_i^{(f)})$$

2. Introduce a Majorana representation for impurities:

$$\{f_{j\alpha}^{+}; f_{j\alpha}\} \longrightarrow \{\eta_j^{\mu}\}_{\mu=0,1,2,3}$$

then $\vec{S}_j^{(f)} + \vec{\tau}_i^{(f)} = -\frac{i}{2} \, \vec{\eta}_j \wedge \vec{\eta}_j$ together with $\vec{\sigma} = -\frac{i}{2} \, \vec{\sigma} \wedge \vec{\sigma}$ for the electrons yields

$$V' = -\frac{1}{8} \sum_{\substack{jm \\ \alpha\beta}} c_{j\alpha m}^{+} c_{j\beta m} (\vec{\sigma} \wedge \vec{\sigma})_{\alpha\beta} \cdot (\vec{\eta}_j \wedge \vec{\eta}_j)$$

3. H_0 being diagonal in the Fourier transformed operators, introduce of Fourier transformation.
4. Introduce an expansion of the partition function Z in powers of J and the cluster expansion for $log Z$. It gives

$$\begin{cases} Z &= \frac{1}{2^L}(\mathrm{Tr}e^{-\beta H_0})(\sum_n J^n U_n) \\ K_0 &= -tL \sum_{kam} c^+_{kam} c_{kam} \delta_{k,0} - \sum_m \tilde{\mu}_m N_m \end{cases}$$

$$(\tilde{\mu}_m = \mu_m - t)$$

$$\frac{1}{\beta} \ln Z = \frac{1}{\beta} \ln \left[\prod_m \frac{(1 + \tilde{z}_m e^{\beta t})}{(1 + \tilde{z}_m)} \right] + \\ + \frac{1}{\beta} \ln [Z_0(t = 0; \{\tilde{\mu}_m\})] + \frac{1}{\beta} \sum_{n=1}^{\infty} J^n W_n$$

with

$$W_n(L) = (-1)^n \int_0^\beta \mathrm{d}\tau_1 \ldots \int_0^\beta \mathrm{d}\tau_n \langle V'(\tau_1) \ldots V'(\tau_n) \rangle_{K_0}^{(c)}$$

Using Wick's theorem W_n is expressed as a finite sum involving propagators $\langle cc \rangle_{K_0}$, $\langle \eta\eta \rangle_{K_0}$. Computing explicitly the propagators, one observes that in the limit $L \to \infty$, all contributions with some $k = 0$ gives zero, while all contributions with all $k \neq 0$ are independent of t: we can thus put $t = 0$. Assuming it is possible to permute the limit $L \to \infty$, with the sum over n yields

$$\frac{1}{\beta} \ln Z(\beta, \{\mu_m\}; t, J) =$$

$$= -2LMt\theta(t) + \frac{1}{\beta} \left[\ln Z_0(\beta, \{\mu_m\}; t = 0, J) + \sum_{n=1}^{\infty} J^n W_n(t = 0) \right] =$$

$$= -2LMt\theta(t) + \frac{1}{\beta} [\ln Z(\beta, \{\mu_m\}; t = 0, J)]$$

4. Exact Ground State Energy and Wave Functions in the Limit $J \to \infty$ (with t Negative and Finite)

We shall restrict our attention to the case where the total electron density is less than one:

$$\sum_m N_m = N < L. \tag{16}$$

Let us recall that

$$H = H_0 + JV, \tag{17}$$

acting on the subspace $\bar{\bar{H}}$.

- Let $\mathcal{H}_{\text{min}} \subset \bar{\mathcal{H}}$, the subspace of ground states for JV, then

$$\forall\, \Psi \notin \mathcal{H}_{\text{min}}, \quad \Psi_{\text{min}} \in \mathcal{H}_{\text{min}}$$
$$\langle \Psi | H | \Psi \rangle - \langle \Psi_{\text{min}} | H | \Psi_{\text{min}} \rangle \stackrel{J \to \infty}{\longrightarrow} \infty$$

Therefore in the limit $J \to \infty$ the ground states of H are in $\mathcal{H}_{\text{min}}$, and we just need to study

$$\langle \Psi_{\text{min}} | H_0 | \Psi_{\text{min}} \rangle \quad \text{with } \Psi_{\text{min}} \in \mathcal{H}_{\text{min}},$$

i.e. the operator $P_{\text{min}} H_0 P_{\text{min}}$.

Since we have considered $N < L$, it follows from Sec. 2 that $\mathcal{H}_{\text{min}}$ is the subspace which contains at most one electron at each lattice site, coupled to the impurity to form a singlet.

- One can then construct an isomorphism $\bar{\mathcal{H}} \to \tilde{\mathcal{H}}$ between our system (with $P_{\text{min}} H_0 P_{\text{min}}$) and a new system of spin-1/2 fermions interacting with spinless bosons, with *exactly* 1 particle at each site, i.e. we have creation and annihilation operators for spin-1/2 fermions and spinless bosons $(F^+_{j\alpha}, F_{j\alpha})$, $\alpha \in \{\uparrow, \downarrow\}$, (B^+_{jm}, B_{jm}), $m \in \{1, 2, \ldots, M\}$ and the mapping

$$\begin{array}{lcl} f^+_{j\alpha}|0\rangle & \mapsto & F^+_{j\alpha}|0\rangle \\ \frac{1}{\sqrt{2}}(c^+_{j\uparrow m} f^+_{j\downarrow} - c^+_{j\downarrow m} f^+_{j\uparrow})|0\rangle & \mapsto & B^+_{jm}|0\rangle \end{array}$$

Introducing the Hamiltonian for the new system:

$$\tilde{H} = \frac{1}{2}|t| \sum_{\substack{ij \\ m\alpha}} F_{i\alpha} B^+_{im} P_{\text{G}} F^+_{j\alpha} B_{jm} - |t| P_{\text{G}} N P_{\text{G}}, \tag{18}$$

where P_{G} is the Gutzwiller operator projecting on the subspace with *exactly* 1 particle at each site, we have:

$$\langle \alpha | P_{\text{min}} H_0 P_{\text{min}} | \beta \rangle = \langle \tilde{\alpha} | P_{\text{G}} \tilde{H} P_{\text{G}} | \tilde{\beta} \rangle \tag{19}$$

Since the first term of $\tilde{H}$ is definite positive the ground state of the multichannel Kondo-lattice model will be defined be those $\tilde{\Psi}$ such that

$$\tilde{H} | \tilde{\Psi} \rangle = -|t|\, N\, | \tilde{\Psi} \rangle \tag{20}$$

and the ground state energy is thus

$$E_{\text{GS}} = -|t|\, N. \tag{21}$$

a) If $N \leq L-2$, one obtains the solution of Eq. (20) expressed in the following way:
Let Y be any subset of the lattice $\Lambda = \{1, 2, \ldots, L\}$ with cardinality $|Y| = L - N - 2$,

$$\underline{\sigma}_Y = \{\sigma_y\}_{y \in Y}, \qquad \sigma_y \in \{\uparrow, \downarrow\}$$

and

$$|\tilde{\Psi}(Y; \underline{\sigma}_Y)\rangle = \sum_{ij \notin Y} \sum{}^{*} F^+_{i\uparrow} F^+_{j\downarrow} \prod_{m=1}^{M} \prod_{x \in X_m} B^+_{xm} \prod_{y \in Y} F^+_{y\sigma_y} |0\rangle$$

where $\sum^*$ denotes the sum over all partitions of $\Lambda \setminus \{Y, i, j\}$ into M subsets, i.e.

$$\Lambda \setminus \{Y, i, j\} = \bigcup_{m=1}^{M} X_m.$$

Then $|\tilde{\Psi}(Y; \underline{\sigma}_Y)\rangle$ is solution of Eq. (20).
b) If $N = L - 1$, then

$$|\tilde{\Psi}_\alpha\rangle = \sum_{i=1}^{L} \sum_{\Lambda \setminus i = \cup X_m} F^+_{i\alpha} \prod_{m=1}^{M} \prod_{x \in X_m} B^+_{xm} |0\rangle$$

is solution of Eq. (20).

5. Concluding Remarks

If we impose periodic boundary conditions, it is possible to express the ground states wave functions of the multichannel Kondo-lattice model in the form of Jastrow product. The ground state energy is $-|t|N$ and highly degenerate.

Acknowledgments

It is my pleasure to express my gratitude to Dr. Ramzi Khuri who took an active part in organising the "Deng Feng Wang Memorial Conference".

References

[1] D.F. Wang, C. Gruber: Invariants of the 1D r^{-2} $t-J$ model, Phys. Rev. B **49**, 15712 (1994);
C. Gruber, D.F. Wang: Exact result of the one dimensional r^{-2} supersymmetric $t-J$ model without translational invariance, Phys. Rev. B **50**, 31031 (1994);
C. Gruber, D.F. Wang: One dimensional lattice models of electrons with r^{-2} hopping and exchange, Proceedings of the satellite meeting of Statphys 19, "Statistical models, Yang-Baxter equation and related topics", edited by F.Y. Wu and M. L. Ge, World Scientific, Singapore (1996), pp. 394–401;
C.-A Piguet, D.F. Wang, C. Gruber: SU(m/n) supersymmetric Calogero-Sutherland model confined in harmonic potential, J. Mod. Phys. B **11** (1997), 1839-1844.

[2] C.-A Piguet, D.F. Wang, C. Gruber: Quantum duality and Bethe ansatze for the Hofstadler problem on the hexagonal lattice, Phys. Lett. A. **209** (1996), 110–114;
D.F. Wang, C. Gruber: A remark on interacting anyons in magnetic field, Phys. Lett. A **201** (1995), 257.

[3] C.-A Piguet, D.F. Wang, C. Gruber: Off-diagonal long-range order and Meissner effect for lattice systems, J. Stat. Phys. 88 (1997), 1363-1369.

[4] D.F. Wang, C. Gruber: Impurity coupled to strongly correlated electron system: Ground state properties, Phys. Rev. B **51** (1995), 4820;
D.F. Wang, C. Gruber: Exactly solvable Kondo-lattice models, Phys. Rev. B **51** (1995), 7476;
D.F. Wang, C. Gruber: On the chiral Hubbard model and the chiral Kondo model, J. Stat. Phys. **82** (1996), 421;
D.F. Wang: On chiral Hubbard model at strong interaction, Proceedings of the satellite meeting of Statphys 19, "Statistical models, Yang-Baxter equation and related topics", edited by F.Y. Wu and M. L. Ge, World Scientific, Singapore (1996), pp. 330–333;
D.F. Wang, C. Gruber: Stability of the insulating phase in the chiral Kondo-lattice model, Phys. Rev. B **53** (1996), 34;
C.-A Piguet, D.F. Wang, C. Gruber: Exact solution of the multichannel Kondo-lattice model with infinite range hopping, J. Low Temp. Phys. **106** (1997), 3.

THE ONE-DIMENSIONAL t-J MODEL WITH LONG RANGE INTERACTION

James T. Liu
Michigan Center for Theoretical Physics
Randall Laboratory
Physics Department
University of Michigan
Ann Arbor, MI 48109–1120
jimliu@umich.edu

How many roads must a man walk down
Before you call him a man?
Yes, 'n' how many seas must a white dove sail
Before she sleeps in the sand?
Yes, 'n' how many times must the cannon balls fly
Before they're forever banned?
The answer, my friend, is blowin' in the wind,
The answer is blowin' in the wind.

—Bob Dylan, 1962

Abstract This brief overview of the one-dimensional t-J model with $1/r^2$ interaction is dedicated to the memory of D. F. Wang, who always had a fascination with such integrable systems.

1. Introduction

Models of strongly correlated electron systems have received much recent interest. In particular, the suggestion of Anderson that the two dimensional Hubbard model contains the essential physics of high temperature supercon-

The Universality of Physics: A Festschrift in Honor of Deng Feng Wang.
Edited by Khuri *et al.* Kluwer Academic/Plenum Publishers, 2001.

ductivity and that its normal state may share the Luttinger-liquid-like feature of one dimensional interacting electron systems [1, 2] has generated renewed activity in the study of such strongly correlated low-dimensional models.

Along the same lines, great progress has been achieved in the investigation of low-dimensional integrable systems. Examples include one-dimensional electron systems with contact interaction [3], the Hubbard model, which was solved by Lieb and Wu [4, 5, 6] using the Bethe-ansatz, and the short range t-J model [7]. Another interesting class of exactly solvable models are the Haldane-Shastry-type models, which have long range $1/r^2$ interactions [8, 9].

D. F. Wang was particularly interested in such long range models, and was fascinated with their properties, and especially with the notion that they remain integrable despite their long range interactions. I have had the wonderful pleasure to have worked with him, and to have shared in his excitement in discovering new features and new models of this sort. Our collaborations have always been stimulating and rewarding, and have led to a series of joint publications [10, 11, 12, 13, 14]. Here I hope to provide a brief glimpse at one of the models that lies at the heart of much of D. F. Wang's work, namely the one-dimensional t-J model with long range interactions.

In general, the t-J model is a lattice model of electrons with interaction and hopping. Each site of the lattice may be either occupied by a spin-up electron, a spin-down electron, or empty. Thus this model has three possible states at each site. The Hamiltonian for the t-J model is given by

$$H = -\tfrac{1}{2}\sum_{i\neq j,\sigma}[c^\dagger_{i\,\sigma}c_{j\,\sigma} + \text{h.c.}] + \tfrac{1}{2}\sum_{i\neq j} J_{ij}P_{ij}, \tag{1}$$

where $\sigma =\uparrow,\downarrow$ and P_{ij} is the spin exchange operator,

$$P_{ij} = c^\dagger_{i\,\downarrow}c_{i\,\uparrow}c^\dagger_{j\,\uparrow}c_{j\,\downarrow} + \cdots . \tag{2}$$

In Eq. (1), t_{ij} is the hopping strength, and J_{ij} is the exchange interaction strength. Furthermore, there is an implicit projection onto single occupancy (which may be relaxed in the case of a Hubbard model).

At this stage, neither the dimensionality of the model nor the lattice structure has been specified. Such features may be encoded in the structure of t_{ij} and J_{ij}. For example, for a one-dimensional model, we simply take the lattice sites i and j to run from 1 to N. The short range (nearest neighbor) model is then given by

$$t_{ij} = t\delta_{i,j+1}, \qquad J_{ij} = J\delta_{i,j+1}, \tag{3}$$

where periodic boundary conditions may be naturally imposed.

In general, many one-dimensional nearest neighbor models may be solved by Bethe ansatz. The idea is as follows. For n particles, one may write the multi-particle wavefunction as

$$|\Phi\rangle = \sum_{x_i} \phi(x_1, x_2, \ldots, x_n) c_{x_1}^\dagger c_{x_2}^\dagger \cdots c_{x_n}^\dagger |0\rangle, \tag{4}$$

where ϕ must have the proper (anti-)symmetry under particle interchange. Assuming only two-body interactions, particles in one dimension can only scatter by passing each other. Thus we take the ansatz

$$\phi(x_1, x_2, \ldots, x_n) = \sum_{\text{permutations}} A(P) e^{ik_i x_{P_i}}, \tag{5}$$

where $\{P_i\}$ is a permutation of $\{i\}$. Thus the Bethe-ansatz state is essentially built out of free particle wavefunctions (reflecting the nature of the problem). Inserting this ansatz into the eigenvalue problem then leads to a set of Bethe-ansatz equations for the weights $A(P)$.

While it appears that this technique only works for contact interactions, perhaps surprisingly the supersymmetric t-J model with $1/r^2$ interaction is also amenable to a Bethe-ansatz-like solution. The supersymmetric model corresponds to taking $J_{ij} = t_{ij}$, indicating an equal interaction strength for both electron-hole hopping and electron-electron exchange. In this limit, bosons (holes) and fermions (electrons) are treated symmetrically, corresponding to $SU(1|2)$ supersymmetry, and the Hamiltonian, (1), reduces simply to the supersymmetric form

$$H = -\tfrac{1}{2} \sum_{i \neq j} J_{ij} \Pi_{ij}, \tag{6}$$

where Π_{ij} is the *graded* permutation operator, interchanging bosons and/or fermions as appropriate.

There is a natural generalization of the $SU(1|2)$ model to arbitrary $SU(m|n)$ supersymmetric models. For example, $SU(0|2)$, or simply $SU(2)$, corresponds to the absence of holes (at half filling), and $SU(2|2)$ to a supersymmetric extended Hubbard model [6, 7]. For nearest neighbor interactions, the latter is solvable by Bethe-ansatz in one dimension [17].

Supersymmetry alone is not sufficient for demonstrating the integrability of the model. In particular, the nature of the interaction itself is important. In contrast to the short range model, $J_{ij} = J\delta_{ij}$, the long range $1/r^2$ model is obtained by taking $J_{ij} = 1/d(i,j)^2$ where

$$d(i,j) = \frac{L}{\pi} \sin\left(\frac{\pi|i-j|}{L}\right) \tag{7}$$

for a uniform lattice with periodic boundary conditions. This trigonometric interaction corresponds to the chord distance between points i and j on a closed ring. This long range $SU(1|2)$ model is simply the supersymmetric t-J model introduced by Kuramoto and Yokoyama [18, 19, 10], and reduces to the ordinary $SU(0|2)$ Haldane-Shastry spin chain [8, 9] in the absence of holes. The generalization to the $SU(2|2)$ extended Hubbard model was investigated in [11].

The general $SU(m|n)$ model with this form of the interaction may be solved using the asymptotic Bethe-ansatz (ABA) [20]. The motivation for this ABA lies in the fact that the scattering remains essentially two-body in nature, even in the presence of long range interactions. The ABA was used in the $SU(1|2)$ supersymmetric t-J model [21] as well as its $SU(1|n)$ generalization [19]. Furthermore, it was proven for $SU(0|2)$ [22] and $SU(1|2)$ [10] that the ABA gives *exact* results, even in the non-asymptotic regime. For $SU(m|n)$, the ABA was discussed in [12].

This model on a uniform lattice also admits a generalization where the closed ring is threaded by magnetic flux [23, 13, 14]. For a model with long range interaction, the flux may be treated by closing the ring using twisted boundary conditions. While non-zero flux destroys the equality between J_{ij} (which is insensitive to flux) and t_{ij} (which sees the flux), integrability is nevertheless maintained [14].

In addition to the inverse trigonometric model of Eq. (7), another long range model may be defined on a nonuniform one-dimensional lattice, given by the roots $r_1, r_2, \dots, r_L$ of the Lth Hermite polynomial $H_L(x)$ [13]. Here the distance function is given simply by

$$d(i,j) = |r_i - r_j|. \tag{8}$$

Although this open chain no longer has momentum as a good quantum number (and does not admit a Bethe-ansatz solution), it nevertheless yields an exactly integrable system [13]. In addition to the original $SU(0|n)$ model of [13], the $SU(1|n)$ generalization was given in [1, 2], and the $SU(2|2)$ case was considered in [27].

While this is but a small aspect of the study of integrable systems, the t-J model with long range interactions and its various generalizations provide an important class of strongly correlated electron systems which admit exact solutions. It was perhaps during the study of this model, and the resulting publication of Ref. [10], that D. F. Wang nurtured and developed his interests in low-dimensional integrable systems. This brief survey has only been able to touch upon a few topics. But hopefully it has given a flavor of the rich structure behind the systems that captured his imagination.

Acknowledgments

None of this work would have been possible on my part without the fruitful collaboration with D. F. Wang. This research was supported in part by DOE Grant DE-FG02-95ER40899 Task G.

References

[1] P. W. Anderson, Science **235**, 1196 (1987).

[2] Lu Yu, Zhao-bin Su and Yan-min Li, Chinese J. of Physics (Taipei) **31**, 579 (1993), and references therein.

[3] C. N. Yang, Phys. Rev. Lett. **19**, 1312 (1967).

[4] E. H. Lieb and F. Y. Wu, Phys. Rev. Lett. **20**, 1445 (1968).

[5] C. N. Yang, Phys. Rev. Lett. **63**, 2144 (1989).

[6] C. N. Yang and S. C. Zhang, Mod. Phys. Lett. B **4**, 759 (1990).

[7] B. Sutherland, Phys. Rev. **B 12**, 3795 (1975).

[8] F. D. M. Haldane, Phys. Rev. Lett. **60**, 635 (1988).

[9] B. S. Shastry, Phys. Rev. Lett. **60**, 639 (1988).

[10] D. F. Wang, J. T. Liu and P. Coleman, Phys. Rev. B **46**, 6639 (1992).

[11] D. F. Wang and J. T. Liu, Phys. Rev. B **54**, 584 (1996).

[12] J. T. Liu and D. F. Wang, Int. J. Mod. Phys. **B 10**, 3685 (1996).

[13] J. T. Liu and D. F. Wang, Phys. Rev. B **55**, R3344 (1997).

[14] J. T. Liu and D. F. Wang, Phys. Rev. B **56**, 2312 (1997).

[15] F. H. L. Eßler, V. E. Korepin and K. Schoutens, Phys. Rev. Lett. **68**, 2960 (1993).

[16] F. H. L. Eßler, V. E. Korepin and K. Schoutens, Phys. Rev. Lett. **70**, 73 (1993).

[17] F. H. L. Eßler, V. E. Korepin, and K. Schoutens, Int. J. of Mod. Phys. B **8**, 3205 (1994).

[18] Y. Kuramoto and H. Yokoyama, Phys. Rev. Lett. **67**, 1338 (1991).

[19] N. Kawakami, Phys. Rev. B **46**, 1005 (1992).

[20] B. Sutherland J. Math. Phys. **12**, 246 (1971); **12**, 251 (1971); Phys. Rev. A **4**, 2019 (1971); **5**, 1372 (1972).

[21] N. Kawakami, Phys. Rev. B **45**, 7525 (1992).

[22] F. D. M. Haldane, Phys. Rev. Lett. **66**, 1529 (1991).

[23] T. Fukui and N. Kawakami, Phys. Rev. Lett. **76**, 4242 (1996).

[24] A. P. Polychronakos, Phys. Rev. Lett. **70**, 2329 (1993).

[25] D. F. Wang and C. Gruber, Phys. Rev. B **49**, 15712 (1994).

[26] C. Gruber and D. F. Wang, Phys. Rev. B **50**, 3103 (1994).

[27] D. F. Wang, Phys. Rev. B **53**, 1685 (1996).

STABILITY AND SYMMETRY BREAKING IN METAL NANOWIRES

Charles A. Stafford
Department of Physics, University of Arizona
1118 East 4th Street, Tucson, AZ 85721

Abstract A linear stability analysis of metal nanowires is performed in the free-electron model using quantum chaos techniques. It is found that the classical instability of a long wire under surface tension can be completely suppressed by electronic shell effects, leading to stable cylindrical configurations whose electrical conductance is a magic number 1, 3, 5, 6,... times the quantum of conductance.

1. Introduction

I had the good fortune to share an office with Deng Feng Wang when we were graduate students at Princeton, and to share the shores of Lac Léman in Switzerland with him as a postdoc some years later. I always enjoyed Wang's wry sense of humor. Like many of his American friends, he did not think me capable of pronouncing "Deng Feng," insisting that I simply call him "Wang," so that is how I knew him. Wang was always concerned with the happiness of his physics friends, not only scientifically, but also in their personal lives. After many years, I finally succumbed to his exhortations to "get yourself a woman," convincing my girlfriend Helen to move to Tucson with me from Germany, when my academic career brought me back to America.

My career has been inextricably tied to that of Deng Feng Wang. As graduate students, we both studied integrable models of one-dimensional many-body systems. My first foray into the field of metallic nanostructures, which has since become the mainstay of my research program, was in collaboration with Wang. We investigated how interactions between electrons in a metallic ring influence the occupation of the conducting subbands of the ring. We found that repulsive interactions increase the spectral weight in the subband with the largest charge velocity, leading to an enhancement of the persistent current induced by an Aharanov-Bohm flux threading the ring at low temperatures [1, 2].

Wang also supervised the master's thesis of Jérôme Bürki at the École Polytechnique Fédérale de Lausanne, and recommended Jérôme to Dionys

The Universality of Physics: A Festschrift in Honor of Deng Feng Wang.
Edited by Khuri *et al.* Kluwer Academic/Plenum Publishers, 2001.

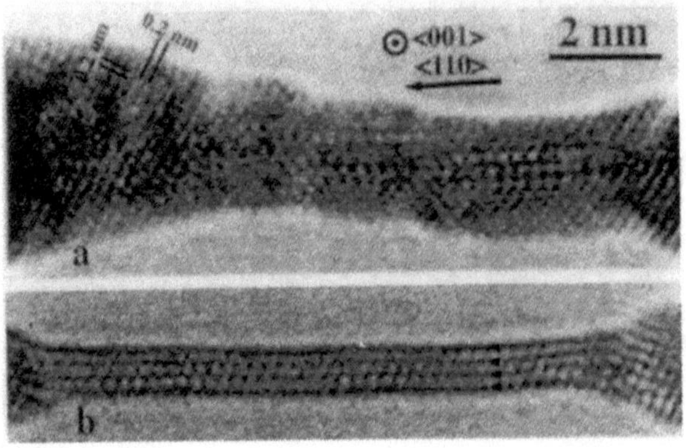

Figure 1. Transmission electron micrographs showing the formation of a gold nanowire, from Ref. [12]: (**a**) an image of Au(001) film with closely spaced nanoholes, the initial stage of the nanowire; (**b**) a nanowire four atoms in diameter, resulting from further electron-beam irradiation.

Baeriswyl, Xenophon Zotos, and myself as a doctoral student to work on a project on metallic nanocohesion funded by the Swiss National Foundation. This was very serendipitous, since that project [3]–[7] was ultimately recognized by the Swiss Physical Society with their 2000 ABB Prize, shared by Dr. Bürki and myself. And this work certainly played no small part in my securing a tenure-track position at the University of Arizona.

I thus think it appropriate, in this volume in honor of Deng Feng Wang, to discuss my most recent work [8] on metallic nanowires. Motivated by Wang's love of mathematical rigor, I will present some of the mathematical details of our stability calculation, which have not been previously published.

2. The Remarkable Stability of the Thinnest Metal Wires

In the past eight years, experimental research on atomically-thin metal wires has burgeoned [9]–[18]. Perhaps the most remarkable feature of metal nanowires is the fact that they are stable at all. Fig. 1 shows electron micrographs illustrating the formation of a gold nanowire [12]. Under electron beam irradiation, the wire becomes ever thinner, until it is but four atoms in diameter. Almost all of the atoms are at the surface, with reduced coordination numbers. The surface energy of such a structure is enormous, yet it is observed to form spontaneously, and to persist almost indefinitely. Even wires one atom thick are found to be stable for days at a time [14, 15].

A cylindrical body longer than its circumference is unstable to breakup under surface tension [19, 20], a phenomenon known as the *Rayleigh instability* (see Fig. 2). How then to explain the durability of long gold nanowires [c.f. Fig. 1(b)], the thinnest of which have been shown [18] to be almost perfectly cylindrical

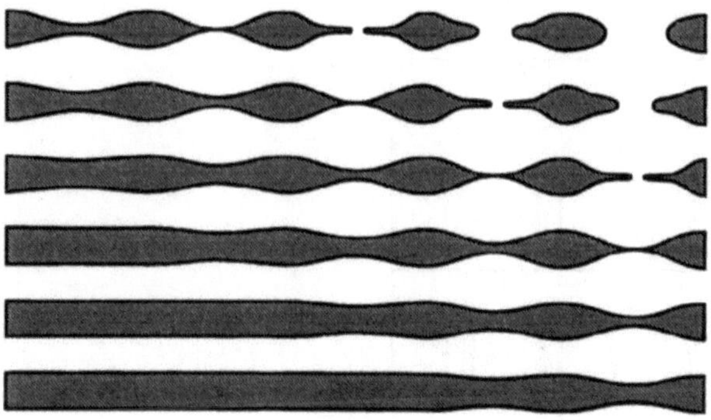

Figure 2. Artist's conception of a propagating Rayleigh instability, from Ref. [21].

in shape? The key is the quantum corrections [8] to the classical stability coefficients.

Only axially-symmetric deformations can lower the surface energy of a cylindrical object, and thus lead to an instability [20]. Any such deformation may be written as a Fourier series

$$R(z) = R_0 + \int_{-\infty}^{\infty} dq\, b(q) e^{iqz}, \tag{1}$$

where $R(z)$ is the radius of the cylinder at z, R_0 is the unperturbed radius, and $b(q)$ is a complex perturbation coefficient. When the wire deforms, the atoms rearrange themselves, but the volume per atom remains essentially constant [6, 22]. The constant-volume constraint leads to the condition

$$b(0) = -\frac{1}{R_0} \int_0^{\infty} dq\, |b(q)|^2. \tag{2}$$

Other physically reasonable constraints are also possible [6], but lead to similar results. Thus the problem is to determine the energetics of a nanowire as a functional of the Fourier coefficients $b(q)$.

3. Free Electron Model

We investigate the simplest possible model [3, 22] for a metal nanowire: a free (conduction) electron gas confined within the wire by Dirichlet boundary conditions. A nanowire is an open system, connected to macroscopic metallic electrodes at each end [9]–[18]. Therefore the change of the grand canonical potential Ω under the perturbation determines its stability. Ω is related to the electronic density of states (DOS) $g(E)$ by

$$\Omega = -\frac{1}{\beta} \int dE\, g(E) \ln\left(1 + e^{-\beta(E-\mu)}\right), \tag{3}$$

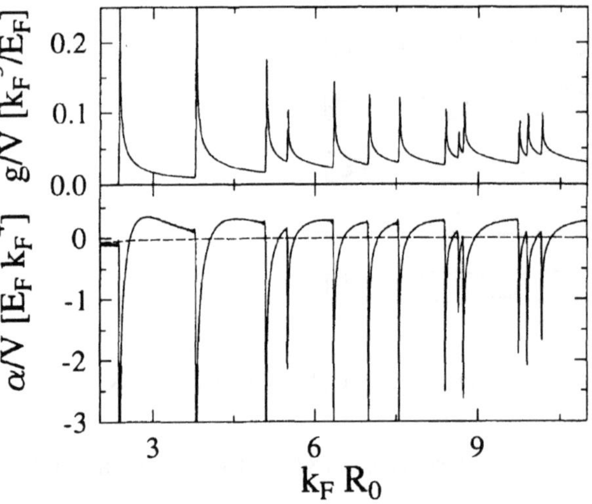

Figure 3. Density of states $g(\varepsilon_F)$ of a cylindrical wire (upper diagram) and stability coefficient α at zero temperature (lower diagram) versus the radius R_0 of the unperturbed wire. The wavevector of the perturbation is $qR_0 = 1$. Dashed curve: Weyl contribution to α.

where β is the inverse temperature and μ is the electrochemical potential of electrons injected into the nanowire from the macroscopic electrodes. The DOS of an open system may be expressed in terms of the scattering matrix as [23]

$$g(E) = \frac{1}{2\pi i} \mathrm{Tr} \left\{ S^\dagger(E) \frac{\partial S}{\partial E} - \mathrm{H.c.} \right\}. \tag{4}$$

This formula is also known as the *Wigner delay*. Note that in Eq. (4), a factor of 2 for spin degeneracy has been included.

Our aim is to expand Ω up to second order in the coefficients $b(q)$ characterizing the deformation. As we will show, this yields

$$\Omega[b] = \Omega[0] + \int_0^\infty dq\, \alpha(q) |b(q)|^2 + \mathcal{O}(b^3), \tag{5}$$

where the stability coefficient $\alpha(q)$ depends implicitly on R_0 and temperature. The change in the grand canonical potential is of second order in b and contributions from deformations with different q decouple. If $\alpha(q)$ is negative for any value of q, then Ω decreases under the deformation and the wire is unstable.

For a translationally-invariant wire, the transverse motion is quantized, leading to the formation of discrete electronic subbands [1, 2]. The total DOS is the sum of the contributions from each subband (see Fig. 3): every subband begins to contribute at a threshold energy equal to the energy of its quantized transverse motion with a sharp spike, falling off smoothly for increasing energy. If the

Fermi energy ε_F lies near one of these sharp peaks, certain small deformations of the wire can dramatically increase the DOS. According to (3), this lowers the grand canonical potential, leading to an instability. On the other hand, if there is no subband threshold sufficiently close to ε_F, we find that the DOS instead decreases with any deformation, implying the existence of stable regions in the intervals between the instabilities associated with the opening of each subband.

4. Weyl Expansion

To examine this picture quantitatively, it is useful to perform a systematic semiclassical expansion [24, 25] of the DOS, $g(E) = \bar{g}(E) + \delta g(E)$, where $\bar{g}$ is a smooth average term, referred to as the Weyl contribution, and $\delta g(E)$ is an oscillatory term, whose average is zero. For the free electron model with Dirichlet boundary conditions, the Weyl term is [25]

$$\bar{g}(E) = E^{-1}\left(\frac{k_E^3 V}{2\pi^2} - \frac{k_E^2 A}{8\pi} + \frac{k_E K}{6\pi^2}\right), \tag{6}$$

where $k_E = \sqrt{2mE}/\hbar$, V is the volume of the wire, A its surface area, and K the integrated mean curvature of its surface. Inserting Eq. (6) into Eq. (3), one finds the following semiclassical expansion at zero temperature:

$$\frac{\Omega}{\varepsilon_F} = -\frac{2k_F^3 V}{15\pi^2} + \frac{k_F^2 A}{16\pi} - \frac{2k_F K}{9\pi^2} + \frac{\delta\Omega}{\varepsilon_F}. \tag{7}$$

One can show [6] that interaction effects are higher order in $\hbar$.

The energy cost of the deformation can be calculated in the Weyl approximation by simple geometrical considerations, and is

$$\begin{aligned}\Delta\bar{\Omega}/\varepsilon_F &= \left(-\frac{8}{15}k_F^3 R_0 + \frac{\pi}{4}k_F^2\right) b(0) \\ &\quad + \int_0^\infty dq \left[-\frac{8k_F^3}{15} + \left(\frac{\pi k_F^2 R_0}{4} - \frac{8k_F}{9}\right) q^2\right] |b(q)|^2. \end{aligned} \tag{8}$$

The constraint (2) can be used to eliminate $b(0)$ in Eq. (8), so that $\Delta\bar{\Omega}$ is quadratic in b, as expected.

5. Trace Formulas

The oscillatory contribution $\delta g(E)$ to the DOS may be approximated as a Feynman sum over classical periodic orbits à la Gutzwiller [24, 25]. Since we are interested in modeling nanowires which possess axial and/or translational symmetries, we can not utilize Gutzwiller's original trace formula [26], which describes systems whose periodic orbits are isolated, but must instead employ a generalization due to Creagh and Littlejohn, describing a system with an f-dimensional Abelian symmetry [27]:

$$\delta g(E) = \frac{2}{\pi\hbar}\frac{1}{(2\pi\hbar)^{f/2}}\sum_{\Gamma}\frac{T_\Gamma V_\Gamma J_\Gamma^{-1/2}}{|\det \tilde{M}_\Gamma - 1|^{1/2}}\cos\left(\frac{S_\Gamma}{\hbar} - \frac{\sigma_\Gamma \pi}{2} - \frac{f\pi}{4}\right), \quad (9)$$

where the sum runs over f-dimensional families Γ of degenerate periodic orbits, T_Γ is the period of an orbit in Γ, V_Γ is the f-dimensional volume spanned by Γ, S_Γ is the action of the orbit, and σ_Γ is a phase shift determined by the singular points along the classical trajectory. The quantity $\tilde{M}$ is the so-called monodromy matrix, characterizing the stability of the orbit with respect to perturbations. It describes as a Poincaré map the linearized motion of small perturbations from the periodic orbit in a surface of section perpendicular to the orbit in phase space: an initial variation of momentum and position in the surface of section $(\delta r, \delta p)$ is related to the mismatch $(\delta r', \delta p')$ after one period by

$$\begin{pmatrix} \delta r' \\ \delta p' \end{pmatrix} = \tilde{M} \begin{pmatrix} \delta r \\ \delta p \end{pmatrix}. \quad (10)$$

Finally, the factor $J_\Gamma = |\det(\partial r'/\partial p)|$.

We shall also need to consider the breaking of continuous symmetries, which is elegantly described in terms of semiclassical perturbation theory [28, 29], wherein the cosine in the trace formula is replaced by

$$\cos(S_\Gamma/\hbar + \theta_\Gamma) \to \mathrm{Re}\left\{e^{i(S_\Gamma/\hbar + \theta_\Gamma)}\left\langle e^{i\Delta S_\Gamma/\hbar}\right\rangle_\Gamma\right\}, \quad (11)$$

where

$$\langle e^{i\Delta S_\Gamma/\hbar}\rangle_\Gamma = V_\Gamma^{-1}\int d\mu(g) e^{i\Delta S_\Gamma(g)/\hbar} \quad (12)$$

is an average over the measure of the broken symmetry group.

For an axially-symmetric three-dimensional nanowire, the periodic orbits (see Fig. 4) occur in one-dimensional families which fit within the circular cross-section of the wire. For the unperturbed cylinder, Eq. (9) yields

$$\delta g(E) = \frac{mL}{\pi\hbar^2}\sum_{w=1}^{\infty}\sum_{v=2w}^{\infty}\frac{f_{vw}L_{vw}}{v^2}\cos(k_E L_{vw} - 3v\pi/2), \quad (13)$$

where $f_{vw} = 1 + \theta(v - 2w)$ counts the discrete symmetry of the orbit under time-reversal and the remaining terms are defined in the caption of Fig. 4. The DOS given by the sum of Eqs. (6) and (13) is plotted in the upper part of Fig. 3.

The deformation (1) breaks the translational symmetry of the wire. The corresponding modulation factor (12) arising in semiclassical perturbation theory

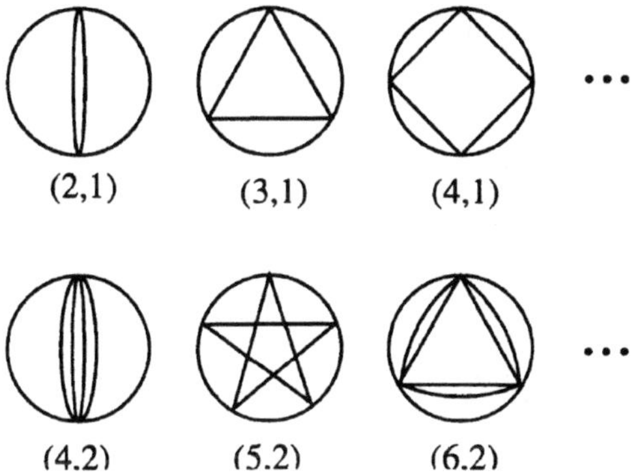

Figure 4. Periodic orbits of an electron in a cross-section of the wire, labeled (v, w), where v is the number of vertices and w the winding number. The length of an orbit is $L_{vw} = vD^* \sin\phi_{vw}$, where $\phi_{vw} = \pi w/v$ is the angle of incidence at a vertex.

is

$$\langle e^{i\Delta S_{vw}(z)/\hbar} \rangle_z = \frac{1}{LR_0} \int_0^L dz\, R(z)\, e^{i\Delta S_{vw}(z)/\hbar}, \tag{14}$$

where

$$\frac{\Delta S_{vw}(z)}{\hbar} = 2v \sin\phi_{vw} k_E \int_{-\infty}^{\infty} dq\, b(q) e^{iqz}. \tag{15}$$

Expanding δg up to second order in $b(q)$ gives

$$\begin{aligned} \Delta\{\delta g(E)\} &= \frac{4m}{\hbar^2} \sum_{w=1}^{\infty} \sum_{v=2w}^{\infty} \frac{f_{vw} \sin\phi_{vw}}{v} \Big[b(0)(\cos\theta_{vw} - k_E L_{vw} \sin\theta_{vw}) \\ &- \frac{k_E L_{vw}}{R_0} \int_0^{\infty} dq\, |b(q)|^2 \left(\sin\theta_{vw} + \frac{k_E L_{vw}}{2} \cos\theta_{vw} \right) \Big], \end{aligned} \tag{16}$$

where $\theta_{vw}(E) = k_E L_{vw} - 3v\pi/2$. The constraint (2) can be used to eliminate $b(0)$ in Eq. (16), so that $\Delta\{\delta g(E)\}$ is also quadratic in b.

6. Stability Analysis

Inserting Eq. (16) into Eq. (3), and performing the energy integral, one obtains the oscillatory quantum correction $\delta\Omega$ up to quadratic order in the Fourier coefficients $b(q)$ of the perturbation. Combining $\delta\Omega$ so determined with Eq. (8) yields Eq. (5), as advertised.

Let us first discuss the stability of a nanowire at zero temperature. Fig. 3 shows the stability coefficient and DOS at the classical stability threshold $qR_0 = 1$ as a function of R_0. The quantum correction destabilizes the wire where the DOS is sharply peaked; but what is more surprising, it *stabilizes* the wire in the intervening intervals.

With these results, we can construct the zero temperature stability diagram for the wire (see Fig. 5(a)). In contrast to Plateau's classical stability analysis [19],

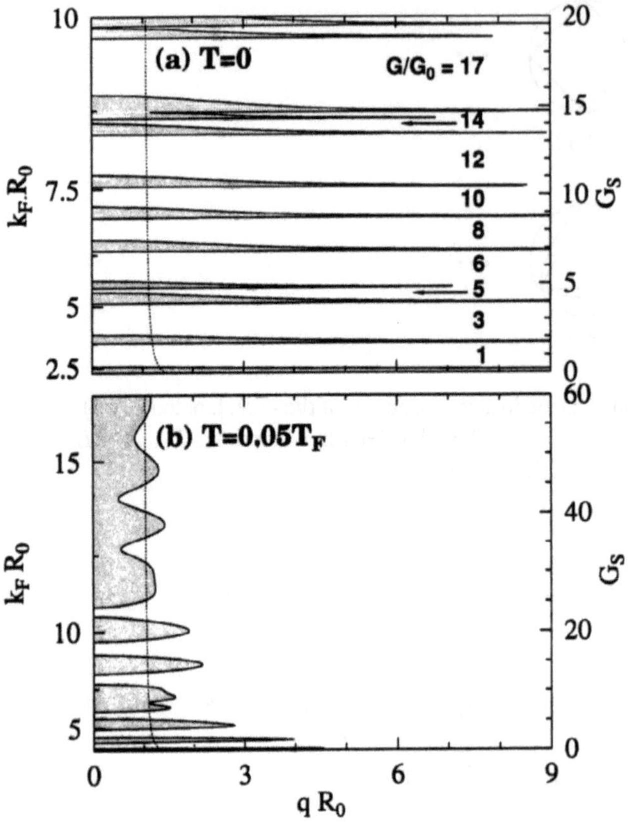

Figure 5. Stability diagram for cylindrical nanowires at two different temperatures. White areas are stable, grey unstable to small perturbations. The quantized electrical conductance values G of the stable configurations are indicated by bold numerals in (**a**), with $G_0 = 2e^2/h$. Right vertical axis: corrected Sharvin conductance G_S. Dotted curve: stability criterion in the Weyl approximation.

an additional quantum length scale arises here, namely the Fermi wavelength λ_F. The stability problem is now determined by two dimensionless parameters: qR_0 and $k_F R_0$. In Fig. 5, regions of instability, where $\alpha(q) < 0$, are shaded grey, while stable regions are shown in white. Note that many of the white regions of stability persist all the way down to $q = 0$, indicating that an infinitely long wire is a true metastable state if its radius lies in one of the windows of stability. The *multistability* of the system, indicated by the alternating stable and unstable stripes, reflects commensurability effects between λ_F and R_0.

The electrical conductance G of a perfect cylindrical nanowire is quantized [30] in units of $G_0 = 2e^2/h$. The quantized conductance values of the stable cylindrical configurations are indicated by bold numerals in Fig. 5(a). For comparison, the right vertical axis of the figure is labeled with the corrected

Sharvin conductance [30]

$$G_S = \left(\frac{k_F R_0}{2}\right)^2 \left(1 - \frac{2}{k_F R_0}\right), \tag{17}$$

which gives a smooth approximation to G/G_0. The conductance values of the stable configurations are somewhat analogous to the *magic numbers* of enhanced stability in atomic nuclei [25] and metal clusters [25, 31]. An important distinction is that the magic numbers in nuclei and clusters refer to the number of fermions in a finite system, while we consider an infinite, open system, with magic numbers describing the number of conducting transverse modes [3] which hold the wire together (the number of conducting modes is approximately equal to the number of atoms which fit within a cross section of the wire). The sequence of magic numbers $G/G_0 = 1, 3, 5, 6, \ldots$ is consistent with the observation of conductance quantizaton in alkali metal nanowires [10, 16, 17].

7. Classical Limit

At zero temperature, the pattern of stable regions separated by unstable stripes shown in Fig. 5(a) continues up to arbitrarilly large radii. However, at any finite temperature T, the quantum oscillations in α are smoothed out, and the classical stability criterion $qR_0 > 1$ is recovered asymptotically for sufficiently large radii. The crossover from the $T = 0$ result to the classical limit occurs when $k_B T \sim E_F(G_0/G)$, i.e., when the thermal energy $k_B T$ is comparable to the average transverse level spacing. Fig. 5(b) shows the stability diagram for $T/T_F = .05$, where $T_F = E_F/k_B$ is the Fermi temperature. One sees that the stability boundary indeed begins to cross over to the classical line $qR_0 = 1$ for $G_S > 20$.

In Fig. 5(b), there are no true metastable configurations with $G_S > 25 \sim T_F/T$, indicating that all thicker wires would be dynamically unstable (like a column of fluid) at this temperature, once the electronic shell effects have been smoothed out. However, $T_F = 3.75 \times 10^4$K in sodium, so multistability from electronic shell effects can be expected to occur in sodium contacts [16, 17] with $G/G_0 \leq 125$ up to at least 300K.

It should be pointed out that thermal averaging is not the only mechanism which can suppress electronic shell effects. Disorder also tends to smooth out the sharply peaked structure in the density of states, so that one can expect a reduction of shell effects when the diameter of the wire exceeds the mean free path. Furthermore, the tendency of the positive ions to order themselves into regular arrays [12, 18] will certainly affect the stability of metallic nanowires. Indeed, pioneering theoretical investigations [32, 33, 34] of the dynamics of nanowires focused exclusively on the arrangement of the ions. Based on the relative importance of electronic shells and crystal structure in metal clusters [31], one would expect electronic shell effects to dominate the energetics of very thin wires, particularly in the alkali metals, with crystal structure becoming increasingly important for thicker wires, and for metals where the bonding is more directional.

8. Conclusions

We have shown that the subband structure [1, 2] of a metal nanowire has a profound effect on its mechanical stability. Our finding that the instability of an atomically-thin metal wire under surface tension can be completely suppressed by quantum-size effects is likely to be important to the eventual commercial development of nanoscale integrated circuits, since it suggests that stable metal interconnects one atom wide are possible. Indeed, linear chains composed of four to seven gold atoms suspended between two gold electrodes, with a conductance $G = G_0$, were found to be stable in the laboratory for hours at a time [14, 15]. Given that such a configuration has an enormous surface energy, its stability is at first sight surprising. However, in our free electron model, we find that an infinitely long wire with a conductance of G_0 is indeed stable with respect to small perturbations.

Finally, let us comment on the dynamical evolution of a nanowire under elongation or compression. Consider stretching a nanowire that is initially in a stable configuration (white areas in Fig. 5). Under elongation, the radius of the wire decreases, so that one moves downward on the stability diagram. When a stability boundary is encountered, it becomes energetically (and dynamically) favorable for the wire to deform spontaneously, until another stable configuration of smaller radius is reached, thus causing the conductance to jump abruptly from one magic number to a smaller one, and conversely under compression. This scenario is consistent with the claim [11] that the structure of a metallic nanowire undergoes a sequence of abrupt changes as a function of elongation or compression. The finite widths of the unstable tongues in Fig. 5 also provides a possible explanation for the hysteresis [11] observed in the conductance as a function of elongation: the critical radius at which the wire's conductance jumps between neighboring magic numbers is different, depending on whether the tongue is approached from above or below, i.e., depending on whether the wire is stretched or compressed.

Acknowledgments

I wish to thank Raymond Goldstein, Hermann Grabert, and especially Frank Kassubek, without whose efforts and insights this work would not have been possible. I also want to thank Helen Giesel for sharing the endless Summer in Arizona with me. This research was supported by NSF Grant DMR0072703 and by an award from Research Corporation.

References

[1] C. A. Stafford and D. F. Wang, Z. Phys. B 103, 323 (1997).

[2] C. A. Stafford and D. F. Wang, Phys. Rev. B 56, R4383 (1997).

[3] C. A. Stafford, D. Baeriswyl, and J. Bürki, Phys. Rev. Lett. **79**, 2863 (1997).

[4] J. Bürki, C. A. Stafford, X. Zotos, and D. Baeriswyl, Phys. Rev. B **60**, 5000 (1999).

[5] J. Bürki and C. A. Stafford, Phys. Rev. Lett. 83, 3342 (1999).

[6] C. A. Stafford, F. Kassubek, J. Bürki, and H. Grabert, Phys. Rev. Lett. **83**, 4836 (1999).

[7] C. A. Stafford, J. Bürki, and D. Baeriswyl, Phys. Rev. Lett. 84, 2548 (2000).

[8] F. Kassubek, C. A. Stafford, H. Grabert, and R. E. Goldstein, Nonlinearity **14**, 167 (2001).

[9] For a review, see *Nanowires*, P. A. Serena and N. Garcia eds. (Kluwer Academic, Dordrecht, 1997).

[10] J. M. Krans, J. M. van Ruitenbeek, V. V. Fisun, I. K. Yanson, and L. J. de Jongh, Nature **375**, 767 (1995).

[11] G. Rubio, N. Agraït, and S. Vieira, Phys. Rev. Lett. **76**, 2302 (1996).

[12] Y. Kondo and K. Takayanagi, Phys. Rev. Lett. **79**, 3455 (1997).

[13] J. L. Costa-Krämer *et al.*, PRB **55**, 5416 (1997).

[14] H. Ohnishi, Y. Kondo, and K. Takayanagi, Nature **395**, 780 (1999).

[15] A. I. Yanson *et al.*, Nature **395**, 783 (1999).

[16] A. I. Yanson, I. K. Yanson, and J. M. van Ruitenbeek, Nature **400**, 144 (1999).

[17] A. I. Yanson, I. K. Yanson, and J. M. van Ruitenbeek, Phys. Rev. Lett. **84**, 5832 (2000).

[18] Y. Kondo and K. Takayanagi, Science **289**, 606 (2000).

[19] J. Plateau, *Statique experimentale et theorique des liquides soumis aux seules forces moleculaires*, (Gautier-Villars, Paris, 1873).

[20] S. Chandrasekhar, *Hydrodynamic and Hydromagnetic Stability* (Dover, NY, 1981) pp 515-74.

[21] T. R. Powers and R. E. Goldstein, Phys. Rev. Lett. **78**, 2555 (1997).

[22] F. Kassubek, C. A. Stafford, and H. Grabert, Phys. Rev. B **59**, 7560 (1999).

[23] R. Dashen, S.-K. Ma, and H. J. Bernstein, Phys. Rev. **187**, 345 (1969).

[24] M. C. Gutzwiller, *Chaos in Classical and Quantum Mechanics* (Springer, NY, 1990).

[25] M. Brack and R. K. Bhaduri, *Semiclassical Physics*. (Addison-Wesley, Reading, MA, 1997).

[26] M. C. Gutzwiller, J. Math. Phys. **12**, 343 (1971).

[27] S. C. Creagh and R. G. Littlejohn, Phys. Rev. A **44**, 836 (1991).

[28] D. Ulmo, M. Grinberg, and S. Tomsovic, Phys. Rev. E **54**, 136 (1996).

[29] S. C. Creagh, Ann. Phys. (N.Y.) **248**, 60 (1996).

[30] J. A. Torres, J. I. Pascual, and J. J. Sáenz, Phys. Rev. B **49**, 16581 (1994).

[31] W. A. de Heer, Rev. Mod. Phys. **65**, 611 (1993).

[32] U. Landman et al., Science **248**, 454 (1990).

[33] T. N. Todorov and A. P. Sutton Phys. Rev. Lett. **70**, 2138 (1993).

[34] M. R. Sørensen, M. Brandbyge, and K. W. Jacobsen, Phys. Rev. B **57**, 3283 (1998).

FROM MAGNETIC FLUX AND INCOMMENSURABILITY TO NMR AND OIL WELLS

Denise E. Freed
Schlumberger Doll Research, Old Quarry Road, Ridgefield, CT 06877

Abstract In this paper I describe some of the work I did as a graduate student about a charged particle in two dimensions subject to a magnetic field, a doubly periodic potential and dissipation. In particular, I discuss how issues of commensurability, incommensurability, and duality symmetries arise in this problem, and I make some connections with work by D. F. Wang. In the second half of this paper, I describe the use of nuclear magnetic resonance (NMR) in the use of oil logging. Finally, I show how questions of commensurability and incommensurability also arise in NMR.

1. Introduction

I met Deng Feng Wang in the graduate administrator Laurel's office. I was the first American graduate student he met. Because I had trouble understanding him when he introduced himself and he was not used to the way an American could mangle his name, he told me it would be easier just to call him Wang. He introduced himself to all our other classmates that way, and the name stuck.

That first year we had many interesting dinners together in the graduate college. I was always fascinated by his unusual way of looking at things. On one of those first few evenings, in trying to make conversation with me, he asked if I liked 'poem-tree'. I thought that was, indeed, a very poetic way of saying peotry. I found it very interesting to hear about his childhood in China, and I was impressed by his strong desire and motivation to 'do something for his country'.

Although Wang and I started out doing quite different research, there turned out to be quite some overlap between our work, especially after we were both no longer at Princeton together. Although, even in our physics, the language we used was quite different, we shared many similar research interests. These included issues to do with magnetic flux and incommensurability, magnetism

The Universality of Physics: A Festschrift in Honor of Deng Feng Wang.
Edited by Khuri *et al.* Kluwer Academic/Plenum Publishers, 2001.

more generally, and Kondo-like scattering problems. We both were especially driven to 1-D solvable models.

In this paper, I will begin by talking about some of my old work on the first of these subjects, that I did while we were both still in graduate school. Then I will describe some of my more recent work in nuclear magnetic resonance for oil logging. Finally, I will show how issues and ideas similar to the questions of flux and incommensurability arise in my current work.

2. Duality and the Hofstadter Spectrum

Suppose you have a charged particle confined to two dimensions that is subject to a 2-D periodic potential and a magnetic field $\vec{B}$ that is perpendicular to the plane. If the potential is very strong, then the system can be described by a hopping problem: To a good approximation, the particle likes to sit at the bottom of the well, and, every so often, it 'tunnels' or 'hops' to an adjacent well.

In the presence of the magnetic field, when the particle hops, it also picks up a phase factor of

$$e^{i2\pi\beta A}, \qquad (1)$$

where A is the area enclosed by the path, in units of the area of the unit cell, and β is the flux per unit cell. β is given by $eBa^2/(2\pi\hbar c)$, where a is the length of the unit cell. Different paths pick up different phases, depending on the area enclosed. These phases interfere, resulting in complicated behavior of the system.

In other words, there are two competing periodicities in this problem. One is the periodic lattice of the potential. The other is a lattice that is defined by the magnetic phase and comes from the periodicity of the phase factor in Eq. (1). This phase factor can be thought of as defining another lattice which has one unit of flux per unit cell. The competition of these two periodicities is characterized by the flux per unit cell β. It turns out that the system has intricate behavior depending on the value of β. When β is rational, so that $\beta = p/q$ with p and q relatively prime integers, it has an unusual dependence on the denominator q. To illustrate this, the energy spectrum as a function of flux per unit cell is plotted in figure (1). This spectrum was first plotted out by Douglas Hofstadter [1] for his Ph.D. thesis, and is known as the Hofstadter spectrum or Hofstdater butterfly, since it looks something like a butterfly.

As can be seen in the figure, if $\beta = 2$, so that there are two units of flux per unit cell, then there are two energy bands. (Strictly speaking, $\beta = 2$ is an exceptional case, and in this case the two bands actually touch, unlike what is shown in the figure.) When $\beta = 1/3$ or $2/3$, there are three energy bands. Similarly, when $\beta = 1/5$, $2/5$, $3/5$ or $4/5$, there are five energy bands, and so on.

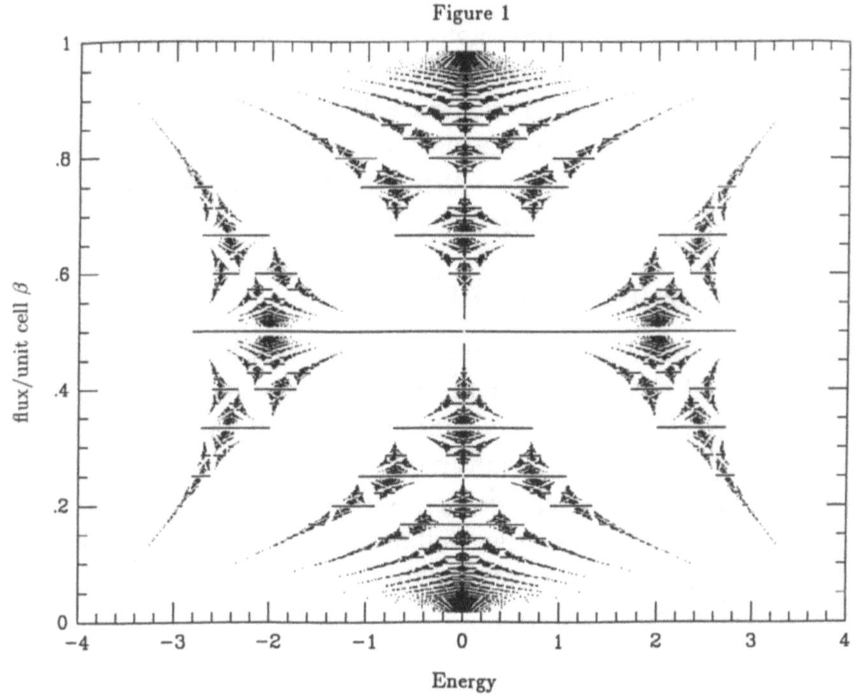

Figure 1. The Hofstadter Spectrum. The black lines denote the allowed energy levels as a function of flux per unit cell, β. Note that when $\beta = p/q$ with p and q relatively prime integers, there are q energy bands.

Another property of the spectrum is that if you take a small part of it, such as looking at only energies less than -1 and looking at flux between $1/3$ and $1/2$, and blow that part up while reshaping it a little, it looks just like the original spectrum. It still has the butterfly pattern running through the center. However, now there is one band when $\beta = 1/3$ and $1/2$ and two bands when $\beta = 2/5$, etc. In this case, then, there are p energy bands when $\beta = p/q$. This property of the spectrum comes from a duality in this model.

To see where this duality comes from, we now consider the case where the potential is very weak. In this case, instead of starting with the particle sitting at the bottom of the well, we can start by looking at the particle in a magnetic field without the potential. The particle will move in a circle, or Landau orbit. The effect of the potential on this orbit is to make the center of the orbit hop by a/β units. Every time the particle feels the effect of the potential, the center of the orbit hops again, so that the positions of the centers of the orbits trace out a new lattice with lattice spacing a/β instead of lattice spacing β. With this new unit cell size, the flux per unit cell becomes $1/\beta$. Thus, we have the same type of hopping problem as in the strong potential case, but this time the flux per unit cell is β instead of $1/\beta$, and this is the origin of the duality.

As a graduate student, I found a new way to calculate the Hofstadter spectrum and to show this duality, using instantons [2]. Later on, Wang looked at a very similar problem. I wish we had had a chance to discuss it at the time. Wang looked at hopping on a hexagonal lattice instead of a square lattice. Among other things, he showed that there was a duality between hopping on the hexagonal lattice and hopping on a triangular lattice, and he found the duality relation between the energy spectra for these two systems [3].

3. The Dissipative Hofstadter Model

To make another, more tenuous connection with Wang's work, I will describe another part of my thesis project. With Curtis Callan, I introduced dissipation to the problem to model the particle's interaction with its environment [4]. This essentially adds a frictional term, with coefficient of friction η, to the equations of motion for the particle. Because we were interested in the effect of dissipation on the quantum mechanical properties of the system, in particular, on the fractal behavior of the energy spectrum, we used the Caldeira-Leggett model to model the friction quantum mechanically [5].

Effectively, this model introduces "interactions" between hopping events. The hopping event at time t_i interacts with the hopping event at time t_j with interaction strength given by

$$\frac{1}{(t_i - t_j)^{-2\alpha \vec{e}_i \cdot \vec{e}_j}}, \tag{2}$$

where $\vec{e}_i$ is the direction of the ith hop, and α can be thought of as the friction per unit cell. It is given by

$$\alpha = \frac{\eta a^2}{2\pi\hbar}. \tag{3}$$

When $\alpha = 1$, this model looks similar to the $1/r^2$ models that Wang worked with [6].

The friction serves to make it more difficult for the particles to hop. If there is no friction (and the flux per unit cell is rational) the particle can hop from well to well and is "delocalized". If the friction is too large, the particle gets stuck in the bottom of a well and is localized. Instead of looking at the energy spectrum for this system, I looked at the phase diagram, which has transitions between localization and different values of mobility when the particle is delocalized. This phase diagram is a function of the friction per unit cell, α, and the flux per unit cell, β. From the simple argument given above, there is a phase transition line for some value of α. Above this line, for greater values of friction, the particle is localized. For smaller values of the friction, the particle can be delocalized. A more detailed analysis shows that this line occurs at $\alpha = 1$ [4].

Without dissipation, we also know that this system has a duality transformation when $\beta \to 1/\beta$. It turns out that with dissipation, this extends to a duality

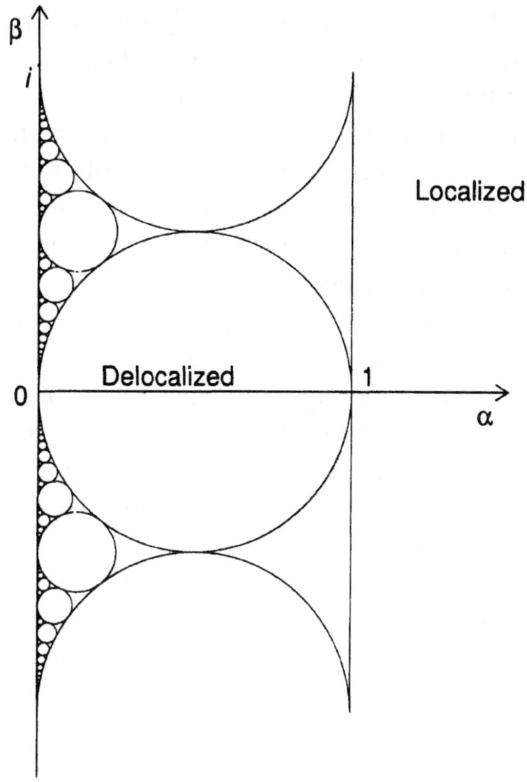

Figure 2. The phase diagram as a function of friction per unit cell α and flux per unit cell β. To the right of the line $\alpha = 1$ the particle is localized. The circles to the left of this line are critical lines. The particle's mobility jumps discontinuously as these lines are crossed.

transformation when $\alpha + i\beta \to 1/(\alpha + i\beta)$ [4]. Under this transformation, the critical line at $\alpha = 1$ is transformed to a circle centered at $\alpha = 1/2$ and $\beta = 1/2$, with radius 1. Inside the circle, the particle is delocalized. The potential is an irrelevant operator for these values of flux and friction, so inside the circle the particle behaves like a free particle subject to flux and dissipation, but no potential. Along the circle, itself, the system is critical, and, in fact, shares some similarities to the anyons with magnetic field [8] and $1/r^2$ t-J model with flux that Wang studied [7]. The energy picks up a phase when the order of the hopping (or, in the case of the anyons and t-J model, the position of the particles) are interchanged.

This system also has a symmetry under the transformation $\beta \to \beta + 1$. This can be seen by looking at the phase factor $e^{i2\pi\beta}$, which remains unchanged under this transformation. The behavior of the mobility under this transformation is a little more complicated.

By repeatedly applying the transformations $\beta \to \beta + 1$ and $\alpha + i\beta \to 1/(\alpha + i\beta)$, we obtain the phase diagram shown in figure (2). To the right of the $\alpha = 1$ line, the particle is localized, effectively stuck in the bottom of a well. To the left of this line, there is a series of circles of radius $1/2$ that are tangent to each other and the lines $\alpha = 1$ and $\alpha = 0$. Inside these circles the particle is delocalized. However, apart from the original circle, inside these circles the mobility can be more complicated than the mobility of the free particle. Also the mobility has a discontinuity as one crosses from the interior of one circle to the next. Between these circles and the line $\alpha = 0$, there is an infinite number of critical circles. Each of these circles is tangent to the line $\alpha = 0$ and two other bigger circles, along with infinitely many smaller circles. Again, inside each of these circles, the particle is delocalized.

The original duality calculation says nothing about the mobility and whether it is localized or delocalized in the wedge-shaped regions between the circles. Since then, I have found that a simple fermionization calculation does show that, for some values of flux and friction in some of the wedge-shaped regions, the particle is still delocalized.

It turns out that every rational value of flux on the $\alpha = 0$ line is tangent to a circle, and every irrational value is tangent to one of the regions that we do not know what is happening in. This meshes well with what we know about the Hofstadter spectrum, which occurs exactly when $\alpha = 0$. For rational values of flux, there are energy bands, so the particle is delocalized. This is consistent with touching a critical circle, in which the particle is always delocalized. For irrational values of flux, the energy spectrum and the behavior of the particle is much more complicated, which we expect could also be the case in some of the wedge-shaped regions. We can see that as the friction is increased, the fractal behavior, characterized in the phase diagram by an infinite number of circles near $\alpha = 0$, is smoothed out until the effects of friction finally dominate for $\alpha > 1$.

4. Oil Logging and NMR

Like Wang, I had to leave my original chosen field of academic physics. I now work at Schlumberger, an oil field services company, in the nuclear magnetic resonance group. I am working on applications of NMR to oil logging.

In this section I will begin by giving a brief description of oil logging. Oil companies drill holes in the ground where they generally expect to find oil. These holes are often several miles deep, but only a few inches wide. Once they have drilled the hole, they need very specific information about whether there is oil and precisely where it is along the length of the borehole. They hire service companies such as Schlumberger, who send tools down the boreholes to determine what is down there. For almost any kind of medical imaging, such as ultrasound or MRI, there is some kind of analogous tool that is used down the borehole. I work on the NMR tool, which is similar to an MRI.

One common misconception about oil wells (that I had before joining Schlumberger) is that the oil is found in huge lakes underground. Instead, the oil is *inside* the rock. Rocks are porous, almost like sponges. If you enlarged a sandstone, for instance, it would look like a whole bunch of grains of sand, cemented together. The spaces between the grains are filled with water, oil or gas. Other rocks can have even more complicated structures. For example, if you magnify limestones, you can often see the remains of microscopic creatures, along with rock crystals and other rock grains. The main goals of oil logging are to look inside the rocks to see how much oil there is, where it is along the borehole, and how easy it is to get it out.

The different tools measure different properties of the oil and rock down the borehole. The most typical quantities of interest include the porosity, the type of fluid, the viscosity and the permeability. The porosity, ϕ, is the ratio of the pore volume to the total volume, so it says how much space could be filled with oil. Fluid typing is used to determine how much of the fluid in the pore space is gas, oil or water, and what type of oil it is. The viscosity, η, is a property of the oil and is a measure of how easily it flows. The permeability k is a property of the rock, and, together with the viscosity, determines how easily the oil will flow out of the rock. In particular, the volume rate of flow per cross-sectional area Q is related to the applied pressure P by

$$Q = \frac{k}{\eta} \nabla P. \tag{4}$$

The NMR tool is of interest because it is the only tool so far that can measure the permeability of the rock. It also measures porosity, as do several other tools, and, if one is clever enough, it can also be used to do rudimentary fluid typing.

The environment in logging is very different from that in the more standard uses of NMR. For 'usual' NMR, such as MRI's or studying the structure of complex molecules like proteins, the subject or sample is put inside a static magnet and rf coils. This geometry provides high, uniform fields, which in turn yield high sensitivity. In the case of MRI's, it is used to produce an image of the subject on the centimeter or millimeter scale. In the case of molecular structure, instead one uses NMR to produce frequency spectra. The different nucleii and the interactions between them produce resonances which provide the information about the molecular structure. One might suspect that we use NMR to image the pore space or to measure the amounts of oil and water by distinguishing their different resonances in a spectrum.

There is a complication, though. For NMR in logging, instead of putting the sample inside the magnet, the magnet and antenna are inside the rock. In other words, in logging there is an 'inside-out' geometry. The first consequence of this is that the signal is not as large, since the fields outside the magnet and antenna are smaller than those inside. Second, the fields outside are inhomogeneous.

The shape of the magnetic field for the Schlumberger tool is a saddle point. Another source of magnetic inhomogeneities is the different susceptibilities of the fluids and the rock, so that even under perfect lab conditions, with the rock inside the magnet, the fields will not be uniform. The inhomogeneities cause the resonances from oil and water to spread out and merge, so we cannot use the spectra to distinguish oil and water. Also, we cannot image the rocks to look at the pore spaces, because the pore size is on the order of 1 to 100 microns, and the imaging only has a resolution of several millimeters at best.

5. Relaxation and Pore Geometry

So, what do the logging NMR tools measure? They measure relaxation, which tells us something about the pore geometry. In this section, I will begin by explaining what relaxation is and what its connection to pore geometery is.

In NMR, there is a static applied field $\vec{B}_0$, which we can assume is pointing along the z-axis. Each hydrogen atom in water and oil has a spin, which in some ways is like a small magnet. In equilibrium, they like to line up in the direction of the applied field, making an equilibrium magnetization which points in the same direction as $\vec{B}_0$. When an rf pulse is applied, the direction of the spins rotates. The pulses can be chosen so that the spins are rotated to the x-y plane and point along the y-axis. Once the spins are in the x-y plane, there are two things that they do. First, they precess around the z-axis, similar to the way a top precesses. They precess with frequency $\omega_0 = i\gamma B_0$, where γ is the gyromagnetic ratio. If we describe the x-y plane by complex numbers, then the direction of the spins as a function of time is given by the phase $e^{i\gamma B_0 t}$.

The second thing the spins in the x-y plane do is to 'relax', meaning that the magnetization in the x-y plane decreases to zero, and the magnetization in the z-direction returns to its equilibrium value. There are several causes of relaxation. For the NMR tool, the largest source of relaxation is the inhomogeneity of the applied field $\vec{B}_0$. Because each spin precesses with frequency $\omega_0 = i\gamma B_0$, if B_0 varies from one point in space to another, the spins in different locations will precess by different amounts. They will eventually point in different directions, or dephase, so that the net magnetization will decay to zero.

The second largest source of relaxation for the spins in the oil or water inside the rock is surface relaxation. When the spins interact with the rock surface, they can change their direction (or effectively disappear), and again the magnetization will decay, with decay time T_2. If we can measure the decay time, it tells us something about how much the surface affects the spins, or the relaxivity ρ of the surface. If we know something about the value of ρ for a particular rock, then the relaxation time T_2 tells us something about the pore geometry [9].

On a more technical level, the spins are diffusing inside a pore, which, for ease of calculation, we take to be a sphere of radius a. When the spin hits the surface of the pore, there is some probability that the spin will disappear. The magnetization then satisfies the Bloch-Torrey equation [10] which is given by

$$\frac{\partial \vec{M}}{\partial t} = D\nabla^2 \vec{M}(\vec{x},t) - \gamma \vec{B}_0(\vec{x},t) \times \vec{M}(x,t) + \text{bulk relaxation.} \tag{5}$$

The first term on the right-hand side describes the diffusion of the particle with diffusion constant D. The second term describes the precession mentioned above. The relaxation at the surface is described by the equation

$$\left(D\hat{n}\cdot\vec{\nabla}M_\perp + \rho M_\perp\right)|_S = 0, \tag{6}$$

where $D\hat{n}\cdot\vec{\nabla}M_\perp$ is the outgoing flux of magnetization, $M_\perp$ is the magnetization in the x-y plane, and S is the surface of the pore.

For many rocks, the spins hit the walls many times before they relax. This is known as the fast diffusion limit, where $\rho a/D << 1$. In this limit, the solution to these equations is fairly simple. The leading behavior is a single exponential decay for each pore, with decay constant T_2 satisfying

$$\frac{1}{T_2} \propto \rho/a, \tag{7}$$

where a is the radius of the pore [11]. More generally, $1/T_2 \propto \rho S/V$, where S is the surface area of a pore and V is its volume. Thus, if we measure the relaxation time in a rock, we end up with a distribution of T_2's which, in turn, gives the pore size distribution inside a rock [9].

For 'well behaved' rocks, the pore size is proportional to the neck size. The 'neck' is the (small) opening between the pores, and is what determines the permeability. In fact, for large classes of rocks, it has been found that the permeability k is related to the relaxation time T_2 and porosity ϕ by [12].

$$k \propto \phi^4 T_2^2. \tag{8}$$

The NMR tool measures this T_2 distribution, which is then used to determine the permeability of the rock. It can also be used to determine how many of the pores are so small that the oil won't be able to flow out of them.

6. CPMG Sequence

There is one catch in this description so far. Recall that the surface relaxation was the second largest cause of the spin's relaxation. The dominant cause of relaxation in the tool's magnetic field is the inhomogeneity of the field. So, in order to measure the relaxation time T_2, we must first remove the effects due to the inhomogeneity of the field B_0. The method for doing this is known as the Car-Purcell-Meiboom-Gill (CPMG) pulse sequence [13]. (In the description of this sequence, we will ignore the diffusion of the spins. Scott Axelrod will address this more complex situation in his paper.)

In the CPMG sequence, the spins start out pointing along the z-axis, aligned with the static field $\vec{B}_0$. Then an rf pulse is applied, which rotates the spins to the x-y plane to lie along the y axis. After a time τ_E, the spins have precessed by an angle $i\gamma B_0 \tau_E$. Because the spins in different locations see a different value of B_0, the spins have dephased and there is no longer any signal from them. The faster spins have gone a farther distance while the slower spins have traveled a smaller distance. Next, another rf pulse is applied, which rotates the spins by 180° about the y-axis. In other words, the spins are reflected about the y-axis, so now the slower spins have the shorter distance back to the starting point along the y-axis, while the faster spins have the larger distance. Consequently, at time $\tau = 2\tau_E$ all the spins line up again along the y-axis. This means that they refocus and form a signal again. This signal is known as an echo. This process of waiting for a time τ_E, applying a 180° pulse, and observing an echo after waiting for time τ_E again is repeated. The successive echoes decay with decay rate $1/T_2$. In this way the effects of the $\vec{B}_0$ inhomogeneities are removed so that the surface relaxation time T_2 can be observed.

7. NMR and Commensurability

In this final section of this paper, we will come full circle and show how the effects of commensurability and incommensurability, reminiscent of that in the Hofstadter spectrum, arise in this NMR system.

So far, I have described how the tool (which was designed well before I joined Schlumberger) is used today. One aspect of my current research involves the fact that CPMG was designed for inhomogeneous fields, but the tool has what can be considered 'grossly' inhomogeneous fields. This has the effect that the 180° pulses are not 'perfect'. This means that if the value of B_0 varies significantly from the nominal value, then the spins are no longer rotated exactly by 180° about the y-axis; both the angle and the axis of rotation vary from these values. The resulting spins' response to the CPMG sequence can be modeled in the following way.

Between pulses, the spins pick up a phase of $e^{i\gamma B_0 \tau_E}$, as before. During the pulse, now only a fraction of the spins will be rotated by 180° about the y-axis. This rotation is equivalent to taking $e^{i\gamma B_0 \tau}$ to $e^{-i\gamma B_0 \tau}$. Another fraction of the spins will be unaffected by the pulse, giving a signal of $e^{i\gamma B_0 \tau}$. The remaining spins in the x-y plane will be rotated back to the z-axis, while fresh spins will be rotated from the z-axis to the y-axis. Repeating this process, we find that

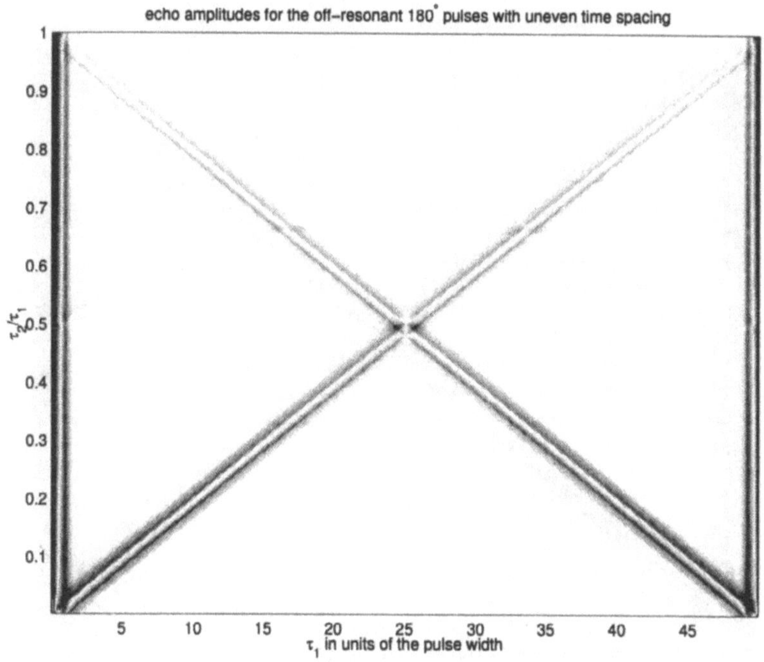

Figure 3. Echoes from off-resonant 180° pulses with uneven time spacings τ_1 and τ_2 as a function of τ_2/τ_1. The x-axis is the time during the interval τ_1 in units of τ_p, where τ_p is the pulse length. Darker colors correspond to stronger intensity echoes.

after many pulses the magnetization has the form

$$M = \sum_{k=-\infty}^{\infty} a_k e^{i\gamma B_0 \tau k}. \tag{9}$$

Now suppose that the intervals $2\tau_E$ between pulses are not perfectly spaced. For example, suppose they alternate, so that the intervals between the pulses are $\tau_1, \tau_2, \tau_1, \tau_2, \ldots$. The phase acquired during time τ_1 is $e^{i\gamma B_0 \tau_1}$, and the phase acquired during τ_2 is $e^{i\gamma B_0 \tau_2}$. The effects of the rotations remain the same. In this case, after many pulses, the magnetization has the form

$$M = \sum_{m,n=0}^{\infty} a_{m,n} e^{i\gamma B_0(\tau_1 n + \tau_2 m)}. \tag{10}$$

In this case, there is again a competition between two periodicities. The signal, once again, depends very much on the commensurability or incommensurability of τ_1 and τ_2 and has intricate behavior depending on p and q, where $\tau_2/\tau_1 = p/q$ with p and q relatively prime integers.

Because, like Wang, I like solvable problems, I found the analytical expression for the long-time behavior of this system. Figure (3) shows the signal

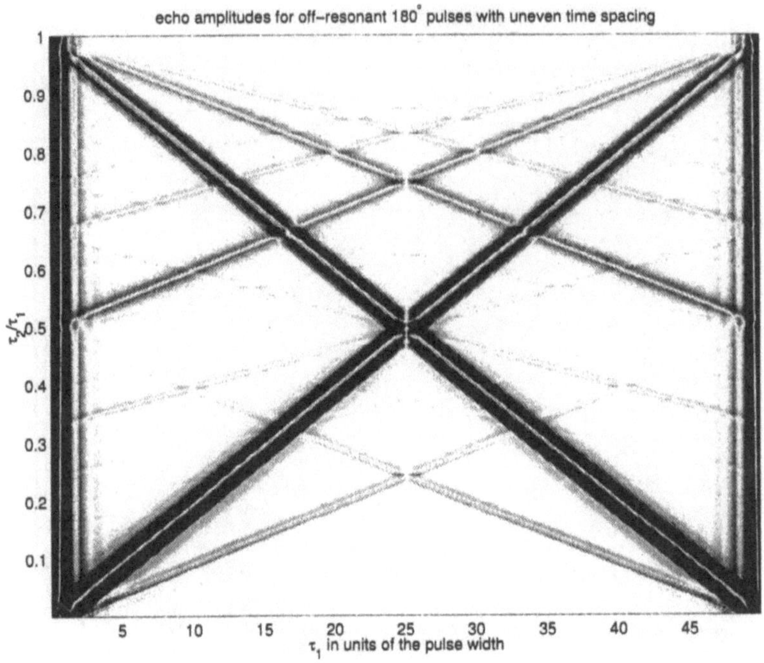

Figure 4. The echoes as in Figure (3), but with the gray scale adjusted to be more sensitive to the weaker echoes and saturated for the stronger echoes.

during the interval τ_1 after many pulses have been applied. The time during the interval is plotted along the x-axis, and the ratio τ_2/τ_1 on the y-axis. The echoes show up as the lines running across the figure.

Note that when $\tau_2/\tau_1 = 1$, there are echoes only at the start and end of the interval, so there is one solid 'band' with no echoes. When $\tau_2/\tau_1 = 1/2$, there are echoes at $t = 0,\ 1/2$ and 1, so now there are two 'bands' with no echoes. When $\tau_2/\tau_1 = 1/3$, there are echoes at $t = 0,\ 1/3,\ 2/3,\ 1$, making three 'bands', and so on. Figure (4) shows a 'blow up' of this plot, where the color scale is much more sensitive to smaller signals and saturates at higher signals. As can be seen, this plot looks fractal again, and the locations of all the X's are very similar to the locations of the 'butterflies' in the Hofstadter spectrum. Even though the X's are not as complex as the butterfly pattern, the competing periodicities in the two systems lead to plots with very similar structure.

In conclusion, although I have switched to a very different field and much of my work is quite applied, even now I have found connections to the physics that Wang and I both shared a love for.

References

[1] D. R. Hofstadter, Phys. Rev. **B14** (1976) 2239.

[2] D. Freed and J. A. Harvey, Phys. Rev **B41** (1990) 11328.

[3] C. A. Piguet, D. F. Wang, and C. Gruber, "Quantum duality and Bethe-Ansatz for the Hofstadter problem on hexagonal lattice".

[4] C. G. Callan and D. Freed, Nucl. Phys. **B374** (1992) 543.

[5] A. O. Caldeira and A. J. Leggett, Physica **A121** (1983) 587; Phys. Rev. Lett. (1981) 211; Ann. Phys. **149** (1983) 374.

[6] C. Gruber, D. F. Wang, "One Dimensional Lattice Models of Electrons with r^{-2} Hopping and Exchange"; D. F. Wang and C. Gruber, "Exactly Sovlable Kondo Lattice Model".

[7] James T. Liu and D. F. Wang, "The $1/r^2$ t-J model in a magnetic field".

[8] D. F. Wang and C. Gruber, "A Remark on Interacting Anyons in a Magnetic Field".

[9] R. L. Kleinberg, Exp. Methods in the Phys. Sciences **35** (1999) 337, and references therein.

[10] H. C. Torrey, Phys. Rev. **104** (1956) 563;

[11] K. R. Brownstein and C. E. Tarr, Phys. Rev. **A19** (1979) 2446-2453.

[12] W. E. Kenyon, P. I. Day, C. Straley, and J. F. Willemsen, Soc. Petrol. Eng. Form. Eval. **3**, (1988) 622-636; erratum: Soc. Petrol. Eng. Form. Eval. **4** (1989) 8.

[13] H. Y. Carr and E. M. Purcell, Phys. Rev. **94** (1954) 630; S. Meiboom and D. Gill, Rev. Sci. Instr. **29** (1958) 688.

GEOMETRY, QUANTUM FIELD THEORY AND NMR

Scott Axelrod
IBM T.J. Watson Research Center
P.O. Box 218
Yorktown Heights, NY 10598
axelrod@us.ibm.com

Abstract We consider the nuclear magnetic resonance (NMR) problem of calculating the Carr-Purcell-Meiboom-Gill (CPMG) echo sequence for spins diffusing in a porous medium in a general magnetic field. We present explicit comparisons between the exact value of the echo sequence and the results predicted asymptotically at the boundaries of the space of parameters of the experiment. We also suggest an exponential term correction to the subleading "boundary" term in the short-time asymptotic formula, which is analogous to the exponential used in quantum field theoretic perturbation theory when going from connected to disconnected Feynman graphs. We discuss briefly how the results presented here answer questions that are analogous to basic questions in geometrical quantum field theory.

1. Introduction

I entered graduate school in the Princeton math department the same year Wang entered in the physics department. (As explained in Denise Freed's talk in the memorial conference, many of our classmates referred to Deng Feng Wang simply as Wang.) Since I was interdisciplinary and had a physics advisor, I hung out with the physics students mostly. Those were very difficult times for most of us and we struggled through together. I will always remember Wang as a great soul and wish I had made more of my opportunity to be influenced by him.

Two incidents come to mind when I think of Wang that illustrate some of the depth of his character. First, I remember the time when we had a birthday party for Wang in one of our dorm rooms. Quite unexpectedly to me, he sang a Chinese song in a beautiful singing voice. It impressed upon me the beauty and power people carry with them even in times of great stress. The other incident happened when a group of us were having lunch together. Wang told of the

The Universality of Physics: A Festschrift in Honor of Deng Feng Wang.
Edited by Khuri *et al.* Kluwer Academic/Plenum Publishers, 2001.

time when he was a student in China and came in one day with his head shaved bald. At the time my hair was covered with overgrown curly hair. He turned to me and offered for the two of us to shave our heads bald together. It was pretty funny and we all laughed, but I do ponder how it might have jolted me out of my own rut if I had been brave enough to follow the lead Wang offered not only in cutting hair but in overcoming his life's struggles before the tragic loss that took him away from us all.

I thought it would be appropriate in honoring Wang, who had a visionary nature and whose career (like my own) branched out from his original roots in and passion for theoretical physics to encompass other areas, to start my talk with a prelude which presents a vision of a connection between some fundamental questions in the area of geometrical quantum field theories (my original field) to the NMR results I will be presenting. We warn the reader in advance that this paper is a combination of fairly general and technical exposition which we thought might be appropriate in this volume.

2. Prelude – Connection with Geometrical Quantum Field Theory and String Theory Picture

2.1. M-Theory

Perhaps the fundamental problem today in basic theoretical physics is to find a "quantum theory of gravity" – a single theory that reproduces the successful predictions of general relativity and quantum field theory in the regimes in which they are each separately applicable. String theory, which is also discussed in Ramzi Khuri's contribution to this volume, is intended to be such a theory. At present, however, it is not really so much a "theory" as it is an evolving attempt to develop a theory based on a starting point that shows much promise at combining the highly geometrical view of the universe found in general relativity with the probabilistic point of view of quantum mechanics. Over the years many different seemingly disparate formulations of string theory have evolved. Recently, however, physicists have observed some remarkable symmetries relating the different formulations and now conjecture they are different asymptotic regimes of a single unifying theory called "M-theory" [1] (where 'M' stands for Mystery, Membrane, Magic, or Mother-of-all-theories). Although they still don't know how to formulate M-theory precisely, they know what many of the parameters involved in it's definition are. They conjecture that, as these parameters go to various different limiting cases, the predictions of M-theory will reproduce the predictions of the various different versions of string theory (or, in one case, something called "11 dimensional supergravity").

It may seem a bit odd that anyone should talk about the limiting case of a theory they don't even know how to formulate, but the situation is not that unfamiliar in quantum field theory. The quantum theory of electromagnetism (QED), for example, has as the basis of its formulation a "Feynman" functional integral over a infinite dimensional space. Unfortunately, this particular functional integral has no real mathematical definition – it is just a formal starting point. However, by applying techniques of "perturbation theory" to extract the asymptotic behavior of a theory as a parameter comes close to a limiting value and a technique called "regularization and renormalization" to cancel out infinite expressions, physicists have for years been predicting the results of experiments very precisely. In this case, they obtain the asymptotic behavior in the regime where the "electromagnetic coupling constant" (which measures the ratio of the strength of the electromagnetic interaction to the size of quantum mechanical effects) comes close to zero.

To reiterate, the full theory of QED (for arbitrary electromagnetic coupling constant) is only formally defined, but the expressions for its asymptotic behavior is rigorously well defined and highly verified by experiment. In the case of string theory, the formal expression that is the starting point is just a bit more nebulous and tantalizing. In the case of M-theory, it is just not known! Nevertheless, the hints we have about M-theory and string theory, which come from theoretical analysis and by looking at the asymptotic regimes where we can, by hook or by crook, arrive at well-defined expression to study, seem to fit together like pieces of a puzzle showing us what's behind the next veil on the face of God.

Two broad question in the subject of M-theory which we will later address in regards to a problem in NMR are:

Q1 What is M-theory? How can one calculate results of M-theory effectively?

Q2 How much of the full M-theory can be determined by it's behavior in the different asymptotic regimes.

2.2. Topological Quantum Field Theory

In 1986, Michael Atiyah proposed a set of axioms as a definition for what he called a "Topological Quantum Field Theory" [2]. This definition is meant to axiomatize the properties of certain kinds of quantum field theories which can be defined on general "space-time manifolds" of some given dimension D; i.e. were the set of possible locations in space and time is a geometrical shape. For example, when $D = 2$, the shape might look like the surface of a doughnut. An important aspect of the theory is to understand what quantum field theories look like on manifolds with boundary and how they behave when the two manifold are glued together along a common boundary. For example, the surface of a pair of pants is a manifold whose boundary is three circles. Two pairs of pants glued together at the waist become a manifold whose boundary is four circles (corresponding to the ends of the cuffs of each of the legs of each of the pairs of pants).

There is a fascinating history of how topological quantum field theories arose and how physicists used them in solving many deep mathematical problems, for example Witten's unification of the theory of knot and 3-manifold invariant as the "physical observables" of certain topological quantum field theories [3]. These theories are wonderful playgrounds for theoretical physicists trying to understand quantum gravity because they do indeed marry quantum mechanics and rich geometry and because, unlike the case of quantum electromagnetism discussed above, they are in fact mathematically well defined. In fact, one can often give very explicit formulas for the full theory rather than just the asymptotic behavior as some parameters go to limiting values.

As for our discussion of M-theory, there is no way that we can do the subject of Topological Quantum Field Theory justice here. So we will again content ourselves with describing two questions that arises and which we will answer later in a much simpler incarnation as questions about NMR.

Q3 How can one calculate perturbation theory for quantum theories defined on manifolds with boundary?

This question arises because physicist's previous experience with perturbation theory for quantum field theories applied predominately to the case when the spacetime manifold had no boundary.

In the case of the NMR problem considered below we will be discussing what a physicist could call Euclidean quantum mechanics on a manifold with boundary which is intimately connected with quantum field theory on a manifold with boundary. Owing to the generality of this paper, we won't delve into trying to make the parallel overly precise.

Q4 How does the perturbation theory in Q3 relate to the full theory?

This question is in the same spirit as the second question [Q2] about M-theory.

3. The NMR Problem

3.1. The CPMG Experiment

In this section we will give an idealization of the CPMG nuclear magnetic resonance experiment due to Carr, Purcell, Meiboom, and Gill [4, 5, 6]. The reader may wish to refer to the article by Denise Freed in this volume for another description of the experiment and a discussion of how it is applied in the context of oil exploration. Although we will be presenting theoretical results here, one motivation of the work presented in this section is in fact to arrive at a better understanding of what properties of the rock surrounding the bore hole of an oil well are reflected in NMR measurements that one can perform in practice. Details for all of the results in this section (apart from section 3.4.3 which presents new results) may be found in [7].

The experimental setup we consider is one in which particles with magnetic moments are diffusing in a region Ω which is a manifold with boundary. In oil field applications for example, the region Ω is (a smooth approximation to) the pore space of a rock. We assume that there is a "background" magnetic field in the region Ω which, for simplicity, we take to point along the z axis. In a typical NMR experiment, in addition to the background magnetic field there is an applied field which is used to control the magnetic moments of the particles. The magnetic moments of the individual particles will both rotate around the direction of the total magnetic field (sum of background and applied fields) as well as "relax" into the direction of the total magnetic field. By measuring the total magnetization $\mathcal{M}(t)$ (the sum of the magnetic moments of all the particles in the system, as a function of time t), one can often deduce important properties of the system under study, for example, an estimate of the distribution of the sizes of pores in a rock. This example was discussed in the contribution to this volume by Denise Freed and is related to the surface to volume ratio appearing in Eq.(24) below.

One difficulty that is encountered in implementing the above is the tendency of the magnetic moments of the individual particles to point in different directions which cancel one another so that the total magnetization is small (and so hard to measure) and also does not reflect simple properties of the system as a whole. In the CPMG experiment, the applied magnetic field is cleverly designed to cause the individual magnetic moments to tend to point in the same direction at certain times. At those times the total magnetization is enhanced and is called the magnetic "echo" since it is viewed as echoing back the magnetic field that was applied. The CPMG experiment is actually a generalization of an experiment due to Hahn [4] who looked at the first echo, often referred to as the "Hahn echo".

At the start of the experiment, a strong field in the z-direction is first applied and held until all of the magnetic moments have relaxed into the z-direction. Then a field "pulse"[1] is applied along the y-axis to cause the magnetic moments of all the particles to rotate until they point along the x-direction. The remainder of the experiment is designed (in the idealization here) so that the magnetic moments of all the particles remain in the $x - y$ plane. So it is meaningful to define $\phi_i(t)$ to be the angle of rotation in the $x - y$ plane from the x-axis to the magnetic moment of the i'th particle. Letting time 0 be the time after the initial setup when all the magnetic moments point along the x-axis, we have $\phi_i(0) = 0$. In the following, we let let $\mathbf{x}_i(t)$ denote the position of the i'th particle at time t.

After the initial setup, the experiment proceeds in four steps. In step 1, no field is applied for time τ while the particles diffuse freely with there magnetic fields rotating about the background magnetic field. The angular velocity with which the magnetic moment of the i'th particle rotates is proportional to $B(\mathbf{x}_i(t))$, the value (at the particle's position) of the z-component of the background field (which we are assuming is the only non-zero component). The constant of proportionality is the gyromagnetic ratio γ. Since $\phi_i(0) = 0$, we have

$$\phi_i(\tau) = \gamma \int_0^\tau B(\mathbf{x}_i(t)). \tag{1}$$

In step 2, a magnetic field pulse is applied causing a 180 degree rotation about the x-axis. This has the effect of changing the sign of the angle ϕ_i. Step 3 is identically to step 1, no field is applied for time τ while the particles diffuse and the magnetic fields rotate about the background field. Finally, in step 4, the total magnetization $\mathcal{M}(2\tau)$ at time 2τ is measured. $\mathcal{M}(2\tau)$ is the first, or Hahn, echo. The n'th echo $\mathcal{M}(2n\tau)$ is obtained by repeating steps 1-4 above repeatedly.

The angle of the magnetic moment for particle i at the Hahn echo time 2τ is

$$\phi_i(2\tau) = \gamma \left[- \int_0^\tau B(\mathbf{x}_i(t)) + \int_t^{2\tau} B(\mathbf{x}_i(t)) \right] \tag{2}$$

Note that if there were no diffusion, i.e. if the particle positions $\mathbf{x}_i(t)$ were constant in time, or if the background magnetic field B were constant in space, the two terms above would cancel exactly. For the higher echoes, there is a similar pairwise cancellation of accumulated angles so that $\phi_i(2n\tau) = 0$ and $\mathcal{M}(2n\tau) = \mathcal{M}(0)$ in case when there is no diffusion or a homogeneous background field.

The combined effects of diffusion and background field inhomogeneity in general give rise to a decrease in magnetization as measure by the *relaxation exponent*:

$$E_n = -\log \left| \frac{\mathcal{M}(2n\tau)}{\mathcal{M}(0)} \right|. \tag{3}$$

The diffusion rate is given by a constant D_0. The size of the magnetic field inhomogeneity is measured by a parameter g which we defined as γ times the root means squared norm of the magnetic field gradient.

3.2. Parameter Space of the System

Three natural length scales of our system can be defined as follows.

- The *diffusion length* $L_D = \sqrt{D_0 \tau}$ equals half the root mean square expected distance a particle will travel between echoes. (The factor of $1/2$ is for consistency with previous conventions.)

- The characteristic size L_S of the region Ω. Given L_S, we define the manifold $\tilde{\Omega}$ by rescaling Ω,

$$\tilde{\Omega} = \{\tilde{\mathbf{x}} = \frac{\mathbf{x}}{L_S}; \mathbf{x} \in \Omega\} \tag{4}$$

  For definiteness, we define L_S to be the unique value so that the volume of $\tilde{\Omega}$ is 1.

- The *phase decoherence length* $L_g = \left(\frac{D_0}{g}\right)^{1/3}$ is a measure of the distance a particle in our system travels before the random effects of diffusion gives rise to appreciable uncertainty in the direction of its magnetic moment.

In the definition above of $\tilde{\Omega}$, the rescaled version of the region Ω, we made use of the dimensionless position coordinate vector $\tilde{x}$. We now also define the dimensionless time coordinate $\tilde{t}$ as well as $\tilde{B}$, $\tilde{D}_0$, and $\tilde{\gamma}$, the dimensionless versions of the field B and the parameters D_0 and γ:

$$\tilde{t} = \frac{t}{\tau}, \qquad \tilde{\mathbf{x}} = \frac{\mathbf{x}}{L_S}, \tag{5}$$

$$\tilde{B}(\tilde{\mathbf{x}}) = \frac{\gamma B(\mathbf{x})}{g L_s}, \tag{6}$$

$$\tilde{D}_0 = \frac{D_0 \tau}{L_S^2} = \left(\frac{L_D}{L_S}\right)^2, \quad \text{and} \tag{7}$$

$$\tilde{\gamma} = \tau g L_s = \left(\frac{L_D}{L_g}\right)^2 \left(\frac{L_S}{L_g}\right), \tag{8}$$

(9)

For each echo number n, the relaxation exponent E_n may be viewed as a function of the dimensionless quantities $\tilde{D}_0$, $\tilde{\gamma}$, $\tilde{B}$, and $\tilde{\Omega}$. For the remainder of the paper we will hold $\tilde{B}$ and $\tilde{\Omega}$ fixed and examine the dependence of the relaxation exponent on the dimensionless diffusion constant $\tilde{D}_0$ and the dimensionless constant $\tilde{\gamma}$ measuring the size of the magnetic field inhomogeneity.

The CPMG experiment is designed just so that (under the idealized model considered here) the relaxation exponent $E_n(\tilde{D}_0, \tilde{\gamma})$ vanishes if either $\tilde{D}_0$ or $\tilde{\gamma}$ does, indicating that there are limiting cases we do know something about. In fact there are three distinct asymptotic regimes, i.e. limits of parameter space, which were studied systematically in [7] building upon a literature of early work. Each case corresponds to one of the three length scales L_D, L_S, or L_g becoming much smaller than the others. In each of the three regimes, a different mechanism is responsible for the strength of the echo $\mathcal{M}(2n\tau)$. In more detail, the three regimes are as follows.

- The *short time* regime occurs when the diffusion length, L_D, is much smaller than the other length scales; i.e. when the ratio $\frac{L_D}{L_S}$ (the fraction of particles hitting the boundary of Ω between echos) and $\frac{L_D}{L_g}$ (the accumulated uncertainty between echoes of magnetic moment phase angles) are very small. In terms of dimensionless parameters, this regime corresponds to the limit when

$$\tilde{D}_0 \ll 1, \tilde{\gamma}^{-2}. \tag{10}$$

This regime is called the *short time* regime because the signal $\mathcal{M}(2n\tau)$ is a perturbation of it's time 0 value $\mathcal{M}(0)$. The asymptotics here is calculated using the fact that the number of interactions one needs to account for between a typical particle and the boundary of Ω or the background magnetic field B is small.

- The *motional averaging regime* occurs when the characteristic system size, L_S, is much smaller than the other scales; i.e. when $\frac{L_D}{L_S}$ (the number of times a particle traverses the region between echoes) and $\frac{L_g}{L_S}$ (the number of traversals taking place before appreciable phase uncertainty) are very large. Equivalently, the requirement is

$$1, \tilde{\gamma} \ll \tilde{D}_0. \tag{11}$$

The term *motionally averaging* is used here because the signal strength comes from averaging what the particle sees over many traversals of the region Ω. The asymptotics in this regime may be calculated using perturbation theory around the lowest eigenmode of the Laplacian on the region Ω (or $\tilde{\Omega}$). (In [7] operator techniques are used which are more tractable than, but equivalent to detailed calculations using eigenmode.)

- The *localization regime* occurs when the decoherence length L_g is by far the smallest. That is, the total dephasing $\frac{L_D}{L_g}$ between echoes is large and the fraction $\frac{L_g}{L_S}$, proportional to the number of particles within a dephasing length of the boundary of Ω is very small. In other words,

$$\tilde{\gamma}^{-2} \ll \tilde{D}_0 \ll \tilde{\gamma}. \tag{12}$$

In this regime, the directions of the magnetic moments for particles in the interior of Ω will have very little coherence (i.e. they will point in uncorrelated directions and so their net contribution to the total magnetization will be negligible). The predominant contribution to the signal in this case comes from the region "localized" near the boundary of Ω where the directions of the magnetic moments is more coherent because the particles tend to move slower near the boundary wall.

3.3. Differential Equation Formulation of NMR Problem

In order to calculate the relaxation exponent $E_n(\tilde{D}_0, \tilde{\gamma})$ as well as its asymptotic behavior in the short time, motionally averaging, and localization regimes, it is convenient to work with the differential equation formulation of the model system we have introduced above. This differential equation is called the Bloch-Torrey equation [8]. In this section we give a full description of the Bloch-Torrey equation written in the rescaled coordinate since it is the best starting point for the analysis necessary to justify the results we will be quoting in the next section.

To write down the rescaled Bloch-Torrey equation explicitly, we first introduce quantities $M_x(\mathbf{x}, t)$ and $M_y(\mathbf{x}, t)$ which denote the local average of the x and y components of the magnetization vectors (lying in the $x - y$ plane) of the particles near $\mathbf{x} \in \Omega$ at time t. Owing to the magnetic field pulses at times $\tau, 3\tau, 5\tau, \ldots$ that rotate all magnetic moments by 180 degrees about the x-axis, we expect M_y to discontinuously change signs at those pulse times. To counter this, we define the function $f(\tilde{t})$ as follows:

$$f(\tilde{t}) = \begin{cases} +1 & \text{if } 4n-1 \leq \tilde{t} < 4n+1 \text{ for some integer } n \\ -1 & \text{if } 4n+1 \leq \tilde{t} < 4n+3 \text{ for some integer } n \end{cases} \tag{13}$$

This definition is just what is required so that the complex valued magnetization function

$$M(\mathbf{x}, t) = M_x(\mathbf{x}, t) + i f(\tilde{t}) M_y(\mathbf{x}, t) \tag{14}$$

will be continuous. The variant of the magnetization function in terms of the dimensionless coordinates is:

$$\tilde{M}(\tilde{\mathbf{x}}, \tilde{t}) = L_S^D M(\mathbf{x}, t), \tag{15}$$

where D is the dimensionality of space for our model[2]

The total magnetization may be written as a simple integral of the magnetization function (or its dimensionless variant):

$$\mathcal{M}(t) = \int_{\mathbf{x}\in\Omega} d^D\mathbf{x}\, M(\mathbf{x}, t) = \int_{\tilde{\mathbf{x}}\in\tilde{\Omega}} d^D\tilde{\mathbf{x}}\, \tilde{M}(\tilde{\mathbf{x}}, \tilde{t}). \tag{16}$$

Using the simplifying assumptions we have made above, the Bloch-Torrey equation for the dimensionless local magnetization is:

$$\frac{d}{d\tilde{t}} \tilde{M}(\tilde{\mathbf{x}}, \tilde{t}) = \tilde{D}_0\, \Delta \tilde{M}(\tilde{\mathbf{x}}, \tilde{t}) - i\tilde{\gamma}\, f(\tilde{t})\, \tilde{B}(\tilde{\mathbf{x}})\, \tilde{M}(\tilde{\mathbf{x}}, \tilde{t}), \qquad \tilde{\mathbf{x}} \in \tilde{\Omega} \tag{17}$$

$$\tag{18}$$

Here Δ is the Laplacian operator acting on functions of the dimensionless position coordinates $\tilde{\mathbf{x}}$.

The boundary condition for this differential equation, assuming that the particles reflect perfectly off the boundary of $\tilde{\Omega}$ with their magnetization unaffected, is that the normal derivative of $\tilde{M}$ at the boundary of $\tilde{\Omega}$ vanishes. For our present purposes, we will make yet another simplifying assumption of spatial uniformity of the magnetization function at time 0. For definiteness, we will normalize so that $\tilde{M}(\tilde{\mathbf{x}}, 0) = 1$. Since L_S is defined just so that the region $\tilde{\Omega}$ has unit volume, this normalization is equivalent to specifying that $\mathcal{M}(0) = 1$.

Equation Eq.(18) together with the boundary and initial conditions described in the previous paragraphs determine the magnetization at all future times.

3.4. Results

3.4.1 Calculating the full theory efficiently. The analog in the NMR context of the first part of question Q1 above about M-theory, is to find and formulate useful NMR experiments and their mathematical models. To "answer" this here, we have simply presented a formulation of the CPMG experiment. The second part of Q2 asks how to calculate results of the theory effectively. For our NMR problem, it turns out that the quantity we are interested, the relaxation exponent for arbitrary $\tilde{D}_0$ and $\tilde{\gamma}$, may be calculated surprisingly efficiently, by truncating the space of all magnetization functions to the space spanned by the first few eigenmodes of the free Laplacian on Ω. For example, in Figure 1, we plot the relaxation exponent for echo number 2 in the case of a simple 1 dimensional problem for which the background magnetic field has constant gradient. The figure shows the relaxation exponent as a function of the dimensionless ratio $\frac{L_D}{L_g}$, when the ratio $\frac{L_g}{L_S}$ is fixed to take the value 0.33 (see [9] to see why this was chosen as an interesting value). Plotted are the approximations calculated by truncating to the first 2 eigemodes and by truncating to the first 3 eigenmodes. The approximation using just 3 (or more) modes converges (to accuracy better than the plot line width) to the exact solution; and the 2 mode approximation already captures the qualitative features of the curve. By contrast, the calculation of the same result by a standard discretization technique, as was done in [9] is much more computationally expensive. The approximation with 3 eigenmodes can in fact be written down analytically and captures the details of the oscillations apparent in Figure 1.

3.4.2 Determining the full theory from it's asymptotic behavior. The NMR analog of question Q2-Q4 in regards to M-theory and topological quantum field theory require us first to characterize the asymptotic behavior of the relaxation exponent in the three limiting regimes describe above (short time, motionally averaging, and localization) and then to see how well knowledge of the behavior in these three limiting cases allows us to determine the exact result. We refer the reader to [7] to see the formulas for the asymptotic behavior in the various regimes as well as their detailed derivations and several plots. We content ourselves here simply to give one plot, Figure 2, which compares the

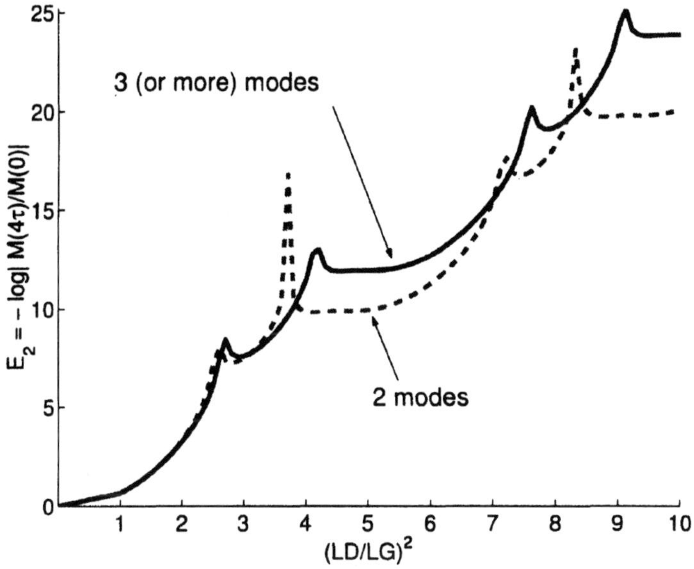

Figure 1. The relaxation exponent for the 1-dimensional system with constant gradient background field for echo number 2 and $(LG/LS) = \left(\frac{\tilde{D}_0}{\tilde{\gamma}}\right)^{1/3} = 0.33$, plotted as a function of $(LD/LG)^2 = \tilde{\gamma}^{2/3}\tilde{D}_0^{1/3}$. Plotted are the approximations calculated using only 2 eigenmodes and with 3 or more eigenmodes. (Any of the latter already converges to the full solution.)

formulas for the asymptotic behaviors to the exact results (calculated using the eigenmode truncation technique of the previous paragraph). In Figure 2 we plot the relaxation exponent for the one dimensional problem with constant gradient background magnetic field for echo number 1 as a function of $\tilde{D}_0$, for fixed $\tilde{\gamma} = 100$. The plot illustrates the validity of the formulas for the asymptotic behaviors in each of the three regimes and shows further that the asymptotic behaviors combined in fact capture the full theory surprisingly well. The fact that the asymptotic behaviors are so successful at capturing the behavior even in the interior of parameter space is no doubt due to the fact that we have chosen to focus on such a simple one dimensional problem. Nevertheless, it does provide an excellent starting point for further exploration.

One other interesting property of Figure 2 is that there is actually an approximate symmetry under the transformation $\tilde{D}_0 \to \tilde{D}_0^{-1}$. More precisely, the short time and motionally averaging regimes can be related using a Poisson summation formula. This is reminiscent of some of the symmetries in string theory that are predecessors of the symmetries that lead to the discovery of M-theory.

3.4.3 More detailed look at short-time perturbation theory vs. full theory. In this sub-section we take a more detailed look at the short time

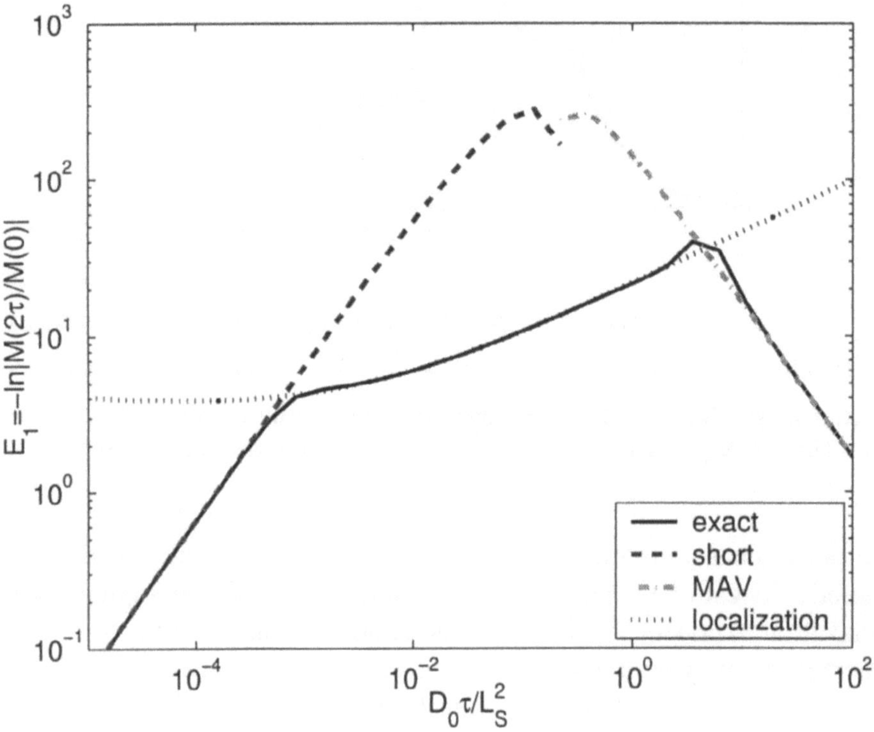

Figure 2. The relaxation exponent for the 1-dimensional system with constant gradient background field for echo number 1 and $\tilde{\gamma} = 100$, plotted as a function $\tilde{D}_0 = \frac{D_0 \tau}{L_S^2} = \left(\frac{L_D}{L_S}\right)^2$. Plotted are the exact relaxation exponent and the asymptotic formulas for the short time, motionally averaging, and localization regimes. Each of the asymptotic formulas provides a good approximation over (more than) its expected range of validity and the asymptotic formulas together provide a surprisingly good approximation over the entire range.

asymptotics of the relaxation exponent and its comparison to the exact result. We will focus on the one dimensional problem already considered above. For this problem, the short time asymptotics of the total magnetization at the n'th echo time $2n\tau$, takes on the following simple form [7],

$$\mathcal{M}(2n\tau)/\mathcal{M}(0) = e^{-E_n(\tilde{D}_0,\tilde{\gamma})} \sim F_0(n, \tilde{\gamma}\sqrt{\tilde{D}_0}) + \sqrt{\tilde{D}_0}\, F_1(n, \tilde{\gamma}\sqrt{\tilde{D}_0}). \quad (19)$$

This equation motivates us to define

$$\alpha = \tilde{\gamma}\sqrt{\tilde{D}_0} = \sqrt{D_0 g^2 \tau^3} = \left(\frac{L_D}{L_g}\right)^3, \quad \text{and} \quad (20)$$

$$\beta = \sqrt{\tilde{D}_0} = \frac{\sqrt{D_0 \tau}}{L_S} = \left(\frac{L_D}{L_S}\right)^2. \quad (21)$$

The series $F_0(n, \alpha)$ may be calculated in a standard way using the technique of "Feynman diagrams" from quantum field theory on a manifold without boundary. A Feynman diagram is a graph representing a kind of possible interaction of the system. To each graph one associates a value (which is given here by a multiple integral over the region Ω). By summing the values associated with all possible diagrams, one obtains the total effect of all interactions. This same result is also obtained by exponentiating the result obtained by summing the values associated to *connected* diagrams only. This is useful because the sum over connected diagram is much easier to calculate since it involves far fewer terms. In fact, for the one-dimensional CPMG problem we are presently considering, there is a single non-trivial connected Feynman diagram whose value is

$$F_0^{connected}(n, \alpha) = -\frac{2n}{3}\alpha^2 = -\frac{2n}{3}\tilde{\gamma}^2 \tilde{D}_0. \quad (22)$$

The exponential of this gives the full series F_0:

$$F_0(n, \alpha) = e^{F_0^{connected}(n,\alpha)} = e^{-\frac{2n}{3}\alpha^2}. \quad (23)$$

The paper [7] discusses how the terms in the series F_1 may be related to calculations in the theory of the short time asymptotics of partial differential equations. These terms may be expressed as multiple integrals over the boundary of Ω (which in the case of the one dimensional problem is simply two points). The integrand in each of these multiple integrals is a differential polynomials in the background magnetic field and the normal vector to Ω^3. The differential polynomials in the expressions for each of the terms in F_1 have coefficients that are universal (i.e. independent of any of the parameters of the system other than the dimension), but are extremely difficult in general to calculate. We claim here that there is in fact an algorithm for expressing the differential polynomial coefficients in terms of multi-dimensional integrals of elementary functions. Unfortunately though, these multi-dimensional integrals are rather intractable.

Prior to the problem of pursuing a more explicit evaluation of the higher order terms in F_1, there is a more basic question to address, namely, to understand the analog of the exponentiation that appears on the right hand side of Eq.(23) which expresses the relation between the sum over connected and disconnected Feynman diagrams. We know of no reference to a similar phenomenon to this exponentiation which would apply to the "boundary term" series F_1, either in the context of NMR theory or in the analogous contexts in topological quantum field theory or string theory.

We close this section with a first primitive stab at exploring whether there might be an analog of the exponentiation on the right of Eq.(23) for the boundary term series F_1. To begin, we write the leading term in F_1 which was calculated in [7]:

$$F_1^{leading}(n,\alpha) = \left[\frac{1}{9}\alpha^2\, C(n)\, \frac{\tilde{S}}{\tilde{V}}\right] \tag{24}$$

Here $\frac{\tilde{S}}{\tilde{V}}$ is the surface to volume ratio of $\tilde{\Omega}$ and $C(n)$ are some easily calculable coefficients which depend on n.

Let us now explore the derivatives of Eq.(19) with respect to β for α held fixed. Rewriting Eq.(19) with the dependence on α and β made explicit, we obtain

$$\mathcal{M}(2n\tau,\alpha,\beta) \sim F_0(n,\alpha) + \beta\, F_1(n,\alpha). \tag{25}$$

Hence,

$$\mathcal{M}(2n\tau,\alpha,0) \sim F_0(n,\alpha), \tag{26}$$

$$\frac{\partial}{\partial\beta}\mathcal{M}(2n\tau,\alpha,\beta=0) \sim F_1(n,\alpha), \quad \text{and} \tag{27}$$

$$\frac{\partial^k}{\partial\beta^k}\mathcal{M}(2n\tau,\alpha,\beta=0) \sim 0 \quad \text{for } k>1. \tag{28}$$

When we calculate $\mathcal{M}(2n\tau,\alpha,\beta=0)$ and $\frac{\partial^k}{\partial\beta^k}\mathcal{M}(2n\tau,\alpha,\beta=0)$ for $k = 2,3$, we find excellent agreement with the right hand sides of Eq.(26) and Eq.(28). For the case $k = 1$ we do not yet know how to calculate the full asymptotic series on the right hand side, however we do know that its leading term is given by Eq.(24). We conjecture that there is a general mechanism to explain how this leading term determines a good approximation of the full series F_1 in a way similar to how Eq.(23) determines the full series from the the connected diagram sum. Since we do not know what that general mechanism is, we will content ourselves here with proposing an approximation to F_1 specific to the CPMG problem we have been studying and which is obtained by modifying

$F_1^{leading}$ by a correction term involving an exponential:

$$F_1^{approx}(n,\alpha) = F_1^{leading}(n,\alpha)\, e^{-0.5n\alpha^2} = F_1^{leading}(n,\alpha)\, e^{-\frac{3}{4}F_1^{connected}(n,\alpha)} \tag{29}$$

The factor 0.5 in the above equation was obtained phenomenologically as a simple fraction giving a good fit. For more involved problems, including the CPMG problem with more complex geometry, we expect that the appropriate generalization of the right hand side of Eq.(29) will be an algebraic formula involving the connected diagrams for the boundaryless problem as well as additional diagrams capturing the effects of the boundary.

In figure 3 we compare the function $F_1^{leading}(n,\alpha)$, $F_1^{approx}(n,\alpha)$, and $F_1^{numerical}(n,\alpha)$, where

$$F_1^{numerical}(n,\alpha) = \frac{\partial}{\partial\beta}\mathcal{M}(2n\tau,\alpha,\beta=0) \tag{30}$$

is calculated by using the technique of Section 3.4.1 and taking derivatives and limits numerically. We see from the figure that $F_1^{leading}$ does indeed capture the leading asymptotics of $F_1^{numerical}$ as required by Eq.(27). We also see that F_1^{approx} provides a much better approximation over the full range of α.

4. Conclusion

In studying a physical system, the following questions are often of fundamental concern:

- What is a useful mathematical model of the system?
- How can one calculate results in thc model effectively?
- What are the limiting cases of the system which are particularly interesting or tractable?
- What is the relationship between the behavior in the limiting cases and the full theory?

We have discussed some answers to these questions in the case when the physical system is a collection of particles moving in a bounded region with background magnetic field. We have focused on a particular idealization of the popular nuclear magnetic resonance experiment measuring the CPMG spin echo sequence.

We also discussed (in an admittedly cursory manner) the above questions in the context of M-theory and topological quantum field theory, both of which have been explored by physicists in trying to better understand the string theory approach to quantum gravity. For M-theory, the particularly pressing problem

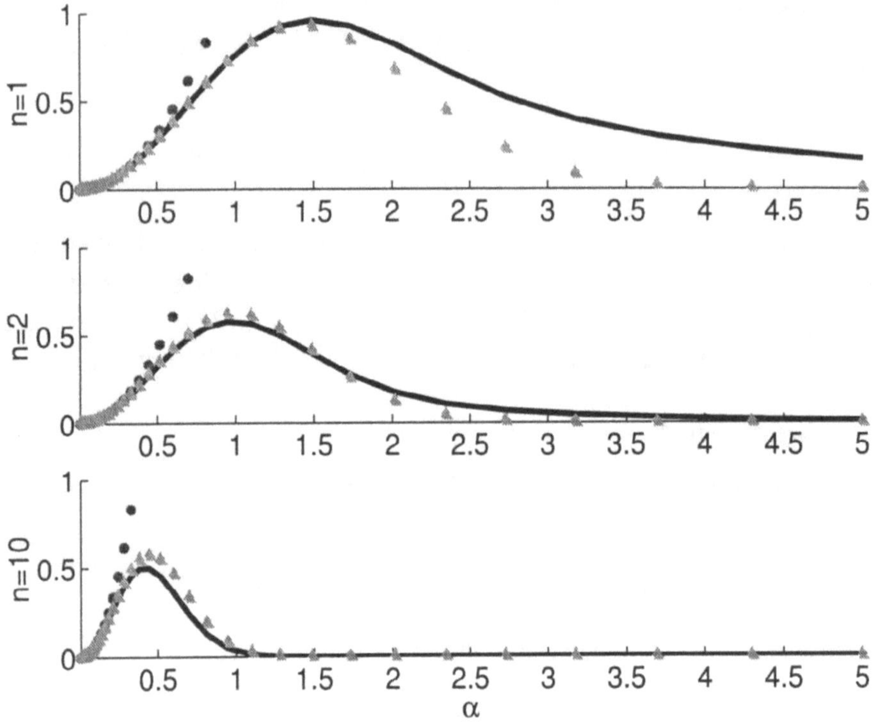

Figure 3. Plots for echo numbers $n = 1, 2,$ and 10 comparing $F_1^{numerical}(n, \alpha)$ (solid black line, from Eq.(30)) $F_1^{leading}(n, \alpha)$ (red circles, from Eq.(24)), and $F_1^{approx}(n, \alpha)$ (green triangles, from Eq.(29)) for the one dimensional CPMG problem with constant gradient background field.

is to find a mathematical model which would say concretely what "M-theory" is, using as a clue the fact that various types of string theories should arise as limiting cases of M-theory. This is a deep and Mysterious problem since we don't even know the basic framework within which to formulate M-theory, although we do know that at some level it will have to incorporate rich geometrical phenomenon such as boundary conditions involving manifolds embedded in other manifolds.

Surely in trying to understand M-theory, physicists will be aided by having a mastery of topological quantum field theories, whose axiomatic framework is much more clear and which are specifically geared toward understanding properties of quantum field theories on manifolds with boundary. Perhaps a more precise understanding of the relationship between quantum field theories on manifolds with boundary and their perturbative limits will be one stepping stone in understanding the relationship between M-theory and it's perturbative limits.

In Section 3, we have considered a much more humble, but more immediately practical, problem in NMR which does involve a physical system constrained to a manifold with boundary. We have provided in Section 3.4.2 a fairly precise relationship between the full theory and its perturbative limits. In Section 3.4.3, we have presented a preliminary exploration of an analog of the exponential relation between connected and disconnected Feynman diagrams. That relation is well known in the study of perturbative quantum field theories on manifolds without boundary, but the analog considered here is applied to a problem with boundaries. Perhaps the formulation here, in terms of a simple, very tractable, model problem could lead to insight into the general problem of relating quantum theories on manifold with boundary to their perturbative limits.

This attempt at relating our little results about NMR to visionary questions in the quantum theory of gravity has been our way of honoring Deng Feng Wang.

Notes

1. By a field pulse, we mean a very strong field which is applied for a very short duration, which we model as a field which is applied instantaneously to cause a discrete rotation of the magnetic moments of all particles.

2. So $D = 3$ for a general three dimension geometry; $D = 2$ for a planar geometry (corresponding, for example, to a three dimensional system where the magnetic field only depends on the x and y coordinates and the region Ω is a cylinder with axis along the z direction); and $D = 1$ for linear geometry (where, for example, the system only depends on the x coordinate).

3. A *differential polynomial* of a collection of functions (such as the background magnetic field and the normal vector field to the boundary of Ω) is a polynomial in the values of the functions and their derivatives.

References

[1] Major developments in M-theory and string theory have been coming out at such a pace that it is hard to give an up to date reference. One recent review available on the web is: M.J.Duff, *State of the unification address*, http://xxx.lanl.gov/abs/hep-th?0012249. We refer the reader to references cited there.

[2] M.F. Atiyah, *New invariants of three and four dimensional manifolds*, in *The mathematical heritage of Hermann Weyl*, Proc. Symp. Pure Math., vol. **48**, ed. R. Wells, Providence, RI, 1988. M.F. Atiyah, *The Geometry and Physics of Knots*, Cambridge U. Press, Cambridge, 1990.

[3] E. Witten, *Quantum field theory and the Jones polynomial*, Comm. Math. Phys., **121**, 351, (1989).

[4] E. L. Hahn, Phys. Rev., **80**, 580, (1950)

[5] H. Y. Carr and E. M. Purcell, Phys. Rev., **94**, 630, (1954)

[6] S. Meiboom and D. Gill Rev. Sci. Instr. **29**, 688 (1958).

[7] S. Axelrod and P.N. Sen, *Nuclear magnetic resonance spin echoes for restricted diffusion in an inhomogeneous field: Methods for asymptotic regimes*, J. Chem. Phys. **114** (2001).

[8] H. C. Torrey, Phys. Rev. **104**, 563 (1956)

[9] P. N. Sen, A. André, S. Axelrod, *Spin echoes of nuclear magnetization diffusing in a constant magnetic field gradient and in a restricted geometry*, J. Chem. Phys., **111**, 6548, (1999).

BLACK HOLES, STRING THEORY AND FUNDAMENTAL PHYSICS

Ramzi R. Khuri

Department of Natural Sciences, Baruch College, CUNY

17 Lexington Ave., NY, NY 10010

khuri@gursey.baruch.cuny.edu

Abstract Quantum aspects of black holes represent an important testing ground for a theory of quantum gravity and ultimately for a theory unifying all four fundamental interactions. The recent success of string theory in reproducing the Bekenstein-Hawking black hole entropy formula provides a link between general relativity and quantum mechanics via thermodynamics and statistical mechanics. We demonstrate new and unexpected links between black holes and polymers and use these links to speculate on the quantum degrees of freedom of black holes.

Keywords: Black holes, quantum gravity, string theory, thermodynamics, polymers.

Prologue: I first met Deng Feng in September, 1986, when we were both newly enrolled students at Princeton. We quickly became close friends. I remember being very impressed by this proud and brilliant, idealistic young man. He was deeply attached to his native country, China, the subject of many of our discussions. He was also a remarkably cultured man, especially so for a 21-year old. He brought with him from China an overpowering ambition to do great things in physics, which he saw as his future contribution to both his country and to the world. I fondly remember the many evenings in which we discussed physics, history, philosophy and one of Deng Feng's favorite subjects, the rights of man. Discussing physics with Deng Feng was truly inspirational: he would make you feel that you were embarking on a fantastic journey, in which you were about to achieve great things. He had an inexhaustible wealth of ideas, and his enthusiasm for science was infectious.

In 1990, Deng Feng returned for a year to China, where he met and married his counterpart and soulmate, the lovely Hu Jing. With Jing at his side, Deng Feng was the picture of self-confidence, producing brilliant, original work in condensed matter physics. Eventually, Deng Feng decided to leave theoretical physics and enter the field of financial mathematics, where again he excelled and quickly produced some outstanding, insightful papers.

The Universality of Physics: A Festschrift in Honor of Deng Feng Wang.
Edited by Khuri *et al.* Kluwer Academic/Plenum Publishers, 2001.

I last saw Deng Feng a little over a week before his tragic accident. I was overjoyed to learn that he would be looking for work in New York. He was full of ideas and possibilities, just as he was when I first met him thirteen years before. His untimely death is a great loss to all of us. Alas, Deng Feng did not get to accomplish all the great things that he wanted to achieve and of which he was capable. Despite his all too short life, Deng Feng had a great influence on those of us fortunate to befriend him. In our hearts and minds, he will live forever.

The standard model of elementary particle physics has been successful in describing three of the four fundamental forces of nature. In the most optimistic scenario, the standard model can be generalized to take the form of a grand unified theory, in which quantum chromodynamics, describing the strong force, and the electroweak theory, unifying the weak interaction with electromagnetism, are synthesized into a single theory in which all three forces have a common origin. The underlying framework of particle physics is quantum mechanics, in which the natural length scale associated with a particle of mass m (such as an elementary particle) is given by the Compton wavelength $\lambda = \hbar/mc$, where $\hbar$ is Planck's constant divided by 2π and c is the speed of light. Scales less than λ are therefore unobservable within the context of the quantum mechanics of this particle.

Quantum mechanics, however, has so far proven unsuccessful in describing the fourth fundamental force, gravitation. The successful theory in this case is that of general relativity, which, however, does not lend itself to a straightforward attempt at quantization. The main problem in such an endeavor is that the divergences associated with trying to quantize gravity cannot be circumvented (or "renormalized") as they are for the strong, weak and electromagnetic forces.

Among the most interesting objects predicted by general relativity are black holes, which represent the endpoint of gravitational collapse. According to relativity, an object of mass m under the influence of only the gravitational force (*ie* neutral with respect to the other three forces) will collapse into a region of spacetime bounded by a surface, the event horizon, beyond which signals cannot be transmitted to an outside observer. The event horizon for the simplest case of a static, spherically symmetric black hole of mass m is located at a radius $R = 2Gm/c^2$, the Schwarzschild radius, from the collapsed matter at the center of the sphere, where G is Newton's constant.

In trying to reconcile general relativity and quantum mechanics, a natural question to ask is whether they have a common domain. This would arise when an elementary particle exhibits features associated with gravitation, such as an event horizon. This may occur provided $\lambda \lesssim R$, which implies that, even within the framework of quantum mechanics, an event horizon for an elementary particle may be observable. Such a condition is equivalent to $m \gtrsim m_P = \sqrt{\hbar c/G} \sim 10^{19} GeV$, the Planck mass, or $\lambda \lesssim l_P = \sqrt{\hbar G/c^3}$, the Planck scale. It is in this domain that one may study a theory that combines quantum mechanics and gravity, the so-called *quantum gravity* (henceforth we use units in which $\hbar = c = 1$).

A problem, however, arises in this comparison, because most black holes are thermal objects, and hence cannot reasonably be identified with pure quantum states such as elementary particles. In fact, in accordance with the laws of *black hole thermodynamics* [1], black holes radiate with a (Hawking) temperature constant over the event horizon and proportional to the surface gravity: $T_H \sim \kappa$. Furthermore, black holes possess an entropy $S = A/4G$, where A is the area of the horizon (the area law), and $\delta A \geq 0$ in black hole processes. So only a black hole with zero area can correspond to a pure state with $S = 0$ such as an elementary particle, while a black hole with nonzero area, and therefore nonzero entropy, corresponds to an *ensemble* of states. A question, then, that can be posed of a theory of quantum gravity is the following: since the basis of ordinary thermodynamics is (quantum) statistical mechanics, can one recover the laws of black hole thermodynamics by the counting of microscopic states? In particular, can one recover the area law from a quantum mechanical entropy arising as the logarithm of the degeneracy of quantum states?

At the present time, string theory, the theory of one-dimensional extended objects, is the only known reasonable candidate theory of quantum gravity. The divergences inherent in trying to quantize point-like gravity seem not to arise in string theory. Furthermore, string theory has the potential to unify all four fundamental forces within a common framework. At an intuitive level, one can see how point-like divergences may be avoided in string theory by considering scattering amplitudes in string theory [2]. Unlike those of field theory, the four-point amplitudes in string theory do not have well-defined vertices at which the interaction can be said to take place, hence no corresponding divergences associated with the zero size of a particle. A simpler way of saying this is that the finite size of the string smooths out the divergence of the point particle.

For the purpose of understanding black hole thermodynamics, an important feature of string theory is that classical solutions [3] may be easily constructed as composites of single-charged fundamental constituents. Identifying these constituents with states in string theory, one can compare the Bekenstein-Hawking entropy obtained from the area of the classical solution to the quantum-mechanical microcanonical counting of ensembles of states [4]. For example, the extremal Reissner-Nordström charged black hole solution of Einstein-Maxwell theory arises in string theory as the composite of four charges, N_1, N_2, N_3 and N_4, normalized to correspond to number operators in string theory. The area law then yields a Bekenstein-Hawking entropy $S_{BH} = 2\pi\sqrt{N_1 N_2 N_3 N_4}$.

The counting of the degeneracy of the states forming this black hole leads to the same quantity $S_{QM} = \ln d(N_i) = S_{BH}$. Even in the black hole picture, this result can be seen to arise from the number of ways in which the various constituents combine. It is straightforward to show [5] that one can write four-centered solutions each with charge N_i of a given species. A black hole with nonzero area is formed when all charges are brought together to the same point. The precise partition function [6] yielding the correct degeneracy $d(N_i) = \exp(S_{BH})$ is obtained provided both bosonic and fermionic excitations of a supersymmetric string-like object along various dimensions are taken into account.

The quantum degeneracy essentially arises from the number of ways of distributing $N_1N_2N_3N_4$ states along 4 bosonic and 4 fermionic degrees of freedom. This yields the above degeneracy and entropy provided the N_k are large. What correspondence then tells us is that the same combinatorics should hold in the black hole picture as well. Another way of saying this is that the horizon forms via string "nucleations" which occur whenever a unit charge of each species combines with unit charges from each of the three other species to form a nonzero horizon "pixel". These nucleations can occur via bosonic or fermionic string degrees of freedom condensing along the four degrees of freedom corresponding to the four "passive" dimensions (which produce no four-dimensional charge upon compactification). The degeneracy is then given by the number of such nucleations that can occur along the four bosonic or four fermionic degrees of freedom. Hence precisely the same combinatoric picture as in the perturbative picture arises directly from the solution, yielding the same degeneracy and entropy as for the quantum states in the zero-coupling limit. Note that this correspondence does not depend on the compactification from which the four-dimensional solutions arises.

The recovery of the area law in a wide variety of contexts in string theory suggests that we have accounted for the microscopic degrees of freedom of the black hole. However, the ensemble of string states on the one hand and the black hole on the other represent two very different objects, so we must try to understand the correspondence between them [7].

Consider a long, self-gravitating string at level N and adiabatically increase the coupling g until the string collapses into a black hole. As noted in [8], the string size at zero coupling (the free string) is initially given by $R_{RW} \sim N^{1/4} l_s$, where l_s is the string scale. The letters "RW" denote "Random Walk", as the free string represents a random walk [9] with $n = N^{1/2}$ steps (or string "bits" [10]). The total length of the string is given by $L = nl_s$. This configuration may be represented as a random walk polymer chain with self-interaction. There are n steps, each of length l, with $\vec{r_i}$ representing the position of the chain after the ith step. Gravitational self-interactions start to become significant once $g \sim g_0 \sim n^{-3/4}$ [8, 11].

This system is described by the generalized Hamiltonian

$$\beta H = \frac{3}{2l}\int_0^L ds \left(\frac{\partial \vec{R}(s)}{\partial s}\right)^2 + g^2 l \int_0^L \int_0^L ds ds' \frac{1}{|\vec{R}(s) - \vec{R}(s')|}, \quad (1)$$

where $\vec{R}(s)$ is the position vector of the chain at arc-length s ($0 \leq s \leq L$). From a Feynman variational procedure for the free energy of the chain [12, 13], It is straightforward to show that the size of the polymer, the average mean square end-to-end distance of the chain, is given by [11]

$$R^2 \simeq \frac{l_s^2}{g^4 n^2}\left(1 - \exp\left(-g^4 n^3\right)\right). \quad (2)$$

For $g << g_0$, $R^2 \sim n l_s^2$, which is the random walk/free string result, while for $g_0 < g < g_c$, $R \sim l_s/(g^2 n)$, which agrees with the calculation of [8, 14].

At zero coupling, $S_0 \sim n$, the number of steps of the random walk. In the intermediate range $g_0 < g < g_c$, the adiabatic increase of the coupling preserves the essential degrees of freedom associated with the string bits. Of course one no longer has a random walk, but up to a factor of order unity, the string bits retain most of their degrees of freedom.

For a black hole whose mass is equal to the excited string state up to a factor of $O(1)$, $S_{BH} = A/4G \sim R_S^2/l_P^2$, where $l_P = g l_s$ is the Planck scale and $R_S \sim GM \sim l_P^2 M \sim l_P^2 n/l_s \sim g^2 n l_s$ is the Schwarzschild radius. We may then rewrite the BH entropy as

$$S_{BH} \sim \frac{n^2 l_P^2}{l_s^2} \sim n^2 g^2. \quad (3)$$

At the critical coupling $g_c \sim n^{-1/2}$, the entropy makes a smooth transition to the Bekenstein-Hawking area law form: $S_{BH} \sim n \sim S_0$. This entropy still represents the degeneracy of a polymer system with n steps (or links). This can be seen as follows: at $g = g_c$, the size of the collapsed polymer string is given by $R \sim R_S \sim l_s$, or the size of one string bit, or one step. In order for n steps of size l_s to fit into a sphere of radius R_S, the number of possible positions to which each step can go must remain a small whole number, p'. Hence the degeneracy again has the random walk form $d \sim p'^n$. Of course, the polymer is no longer a random walk, but the degrees of freedom still remain essentially intact.

Once the transition is complete and the black hole picture prevails, the area of the black hole is given by $A \sim R_S^2 \sim l_s^2 \sim (1/g^2) l_P^2 \sim n l_P^2$. So the horizon can be divided into n "pixels" each of area l_P^2. Once the horizon forms, the degrees of freedom associated with it represent independent quantum states, the points on the horizon are causally disconnected. Again, only a small whole number of possible states, q, is associated with each pixel, so that the total degeneracy is given by $d_{BH} \sim q^n$, with the entropy given by $S = \ln d \sim n \sim S_{BH}$. So essentially the random walk degrees of freedom turn into horizon surface degrees of freedom at the critical transition point. Another way of saying this is that the string bits project their information onto the horizon, in accord with expectations of the holographic principle [15]. So the underlying degrees of freedom of quantum black holes in string theory remain associated with the original, stringy degrees of freedom. This further strengthens the string/black hole correspondence conjecture [7] by implying that in the transition to the strong-coupling limit, it is possible that the quantum string states somehow retain far more of their nature from the perturbative picture than might have been supposed.

This connection between black holes, strings and polymers is very interesting and merits further investigation. Similar links with other soft-matter systems have also been noted in [16], where the area law was recovered for the case of a liquid field theory and where it was argued that the area law contributions to the free energy are primarily responsible for liquid surface tension. The speculation was also made that the area law arises in the context of protein folding.

Connections between physical and biological systems are always exciting. The cases discussed above are especially so since quantum gravity is generally considered too remote to have relevance to other areas of physics, much less other fields of science. In particular, the fascinating possibility arises that mathematical techniques used to study black holes can be useful in understanding biological questions, such as protein dynamics, while methods of polymers physics can potentially shed light on quantum gravity.

Acknowledgments

Research funded by NSF Grant 9900773 and by PSC-CUNY Award 62557 00 31.

References

[1] J. Bekenstein, Lett. Nuov. Cimento **4** (1972) 737; Phys. Rev. **D7** (1973) 2333; Phys. Rev. **D9** (1974) 3292; S. W. Hawking, Nature **248** (1974) 30; Comm. Math. Phys. **43** (1975) 199.

[2] M. B. Green, J. H. Schwarz and E. Witten, *Superstring Theory*, Cambridge University Press, Cambridge (1987).

[3] See M. J. Duff, R. R. Khuri and J. X. Lu, Phys. Rep. **B259** (1995) 213, M. Cvetic and D. Youm, Phys.Rev. **D54** (1996) 2612, M. Cvetic and A. A. Tseytlin, Nucl. Phys. **B477** (1996) 499 and references therein.

[4] A. Strominger and C. Vafa, Phys. Lett. **B379** (1996) 99; J. Maldacena, hep-th/9607235 and references therein; K. Sfetsos and K. Skenderis, Nucl. Phys. **B517** (198) 179; R. Argurio. F. Englert and L. Houart, Phys. Lett. **B426** (1998) 275.

[5] J. Rahmfeld, Phys. Lett. **B372** (1996) 198.

[6] T. M. Apostol, *Introduction to Analytic Number Theory*, Springer Verlag (1976).

[7] L. Susskind, hep-th/9309145; G. T. Horowitz and J. Polchinski, Phys. Rev. **D55** (1997) 6189.

[8] G. T. Horowitz and J. Polchinski, Phys. Rev. **D57** (1998) 2557.

[9] P. Salomonson and B. S. Skagerstam, Nucl. Phys. **B268** (1986) 349; Physica **A158** (1989) 499; D. Mitchell and N. Turok, Phys. Rev. Lett. **58** (1987) 1577; Nucl. Phys. **B294** (1987) 1138.

[10] See C. B. Thorn, hep-th/9607204 and references therein; see also O. Bergman and C. B. Thorn, Nucl. Phys. **B502** (1997) 309.

[11] R. R. Khuri, Phys. Lett. **B470** (1999) 73.

[12] See M. Doi and S. F. Edwards, *The Theory of Polymer Dynamics*, Clarendon Press, Oxford (1986) and references therein.

[13] S. F. Edwards and M. Muthukumar, J. Chem. Phys. **89** (1988) 2435; S. F. Edwards and Y. Chen, J. Phys. **A21** (1988) 2963.

[14] T. Damour and G. Veneziano, Nucl. Phys. **B568** (2000) 93.

[15] G. 't Hooft, gr-qc/9310026; L. Susskind, L. Thorlacius and J. Uglum, Phys. Rev. **D48** (1993) 3743.

[16] D. J. E. Callaway, Phys. Rev. **E53** (1996) 3738.

DESIGN OF A LOW POWER PASSIVE SIGMA DELTA MODULATOR

Feng Chen
Mixed Signal Circuit Laboratory
Texas Instruments
Dallas, Texas 75266

1. Introduction

Conventionally a sampled-data based sigma delta modulator consists of active blocks such as operational amplifiers (opamps) in the loop filter which consume the majority of the power in the modulator. It can be expected that a considerable amount of power could be saved if the active loop filter was replaced with a passive network.

In this paper, a low power passive sigma delta modulator is proposed and implemented. Critical system design issues which affect the signal-to-noise ratio (SNR) performance and power consumption of the modulator will be identified. A design in switched capacitor will be given as an example.

2. System Level Design

A block diagram of 1-bit lowpass passive sigma delta modulator is shown in figure 1. Compared to a lowpass active sigma delta modulator, the major difference is that the loop filter is implemented with a lowpass passive filter. The transfer function of the loop filter is H.

The single-bit quantizer is modeled with a linear model. In addition to the quantization noise E_Q, an additive noise source E_{com} is added to model all additive noise sources at the input of the comparator. Most importantly, an equivalent gain factor G of the comparator is added.

The equivalent gain factor G of the comparator is defined as the ratio of the comparator rms output value to its rms input value. In a passive case, the G is assumed to be constant and is estimated to be roughly $1/|H(f_s/2)|$. It can be on the order of thousands because of the small voltage level at the input of the comparator, while G is unity in the active case.

The Universality of Physics: A Festschrift in Honor of Deng Feng Wang.
Edited by Khuri *et al.* Kluwer Academic/Plenum Publishers, 2001.

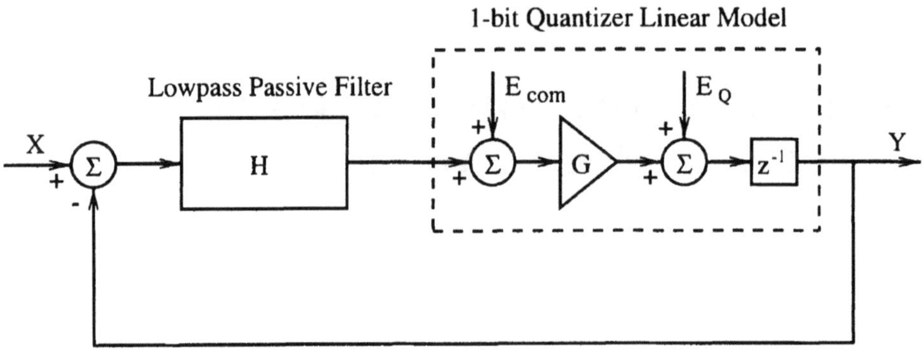

Figure 1. A passive sigma delta modulator with a linear quantizer model.

2.1. Transfer Function

With the above linear model, the output of the modulator is given by:

$$Y = \frac{GHz^{-1}}{1+GHz^{-1}}X + \frac{E_Q z^{-1}}{1+GHz^{-1}} + \frac{Gz^{-1}}{1+GHz^{-1}}E_{com} \tag{1}$$

At low frequency, $GH >> 1$, then the above expression for Y becomes:

$$Y = X + \frac{E_Q}{GH} + \frac{E_{com}}{H} \quad (in\ baseband) \tag{2}$$

In this baseband expression, the first two terms are the same as in the case of an active sigma delta modulator representing the analog input signal and the quantization noise E_Q which has a highpass transfer function of $1/GH$ with attenuation. Furthermore, there is a third term in eqn(2) which is related to the comparator input referred noise E_{com} which has a highpass transfer function of $1/H$ with unity gain. One observation is made here: for a given E_{com}, a passive sigma delta modulator with a large loop gain can always be designed to suppress the quantization noise to a low enough level, so that the E_{com} becomes dominant. In this sense the resolution of the comparator plays a key role in determining the overall SNR of the system. This is one of the essential differences between a passive and an active modulator.

2.2. GH Product

From the above analysis, G can be increased by introducing more attenuation in the loop filter. However, it is the GH product which determines how much the E_Q will be attenuated in the baseband. As pole of the loop filter moves to a frequency which is much lower than the bandedge, the change of H at $f_s/2$ is equal to that of H at bandedge. Therefore GH at bandedge stays unchanged. Assuming that noise at the baseband edge dominates the overall baseband noise

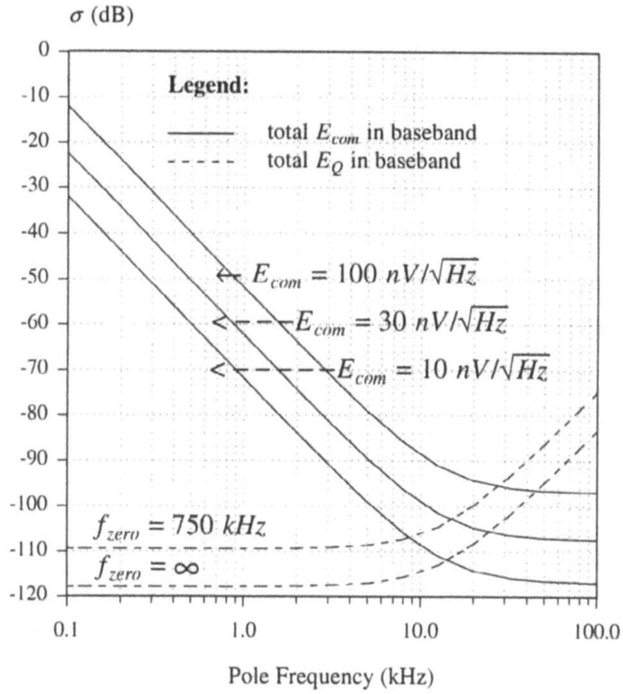

Figure 2. Baseband quantization noise and comparator input noise versus pole frequency (2 identical poles)

power, there will be no further improvement of SNR although the poles are pushed further below the baseband edge. Therefore, it is not necessary to place the poles at a very low frequency.

Another consideration in the placement of poles is the impact of 1/H on E_{com}. The factor 1/H increases in the baseband as the poles move to a lower frequency and hence the input referred E_{com} increases. Eventually E_{com} will dominate the overall noise contribution and determine the modulator's SNR for a given filter order. Therefore, in order to limit the contribution of E_{com}, the poles' frequency should not be too low. Once the filter order is fixed, the loop filter poles can be chosen to satisfy the overall SNR requirement of the system.

2.3. Baseband Noise Estimation

For comparison, the quantization noise σ_q and comparator input referred noise σ_{com} are plotted in figure 2, for an OSR of 256. It can be seen that zero has no impact on the E_{com} contribution, while it does have an impact on that of E_Q. There is an 8.5 dB difference between the two cases with zeros at infinity and 750 kHz. However, as previously discussed, the frequency of the poles affects baseband noise. As the poles move to a lower frequency, the E_{com} contribution increases at a rate of 40 dB/dec while the E_Q contribution

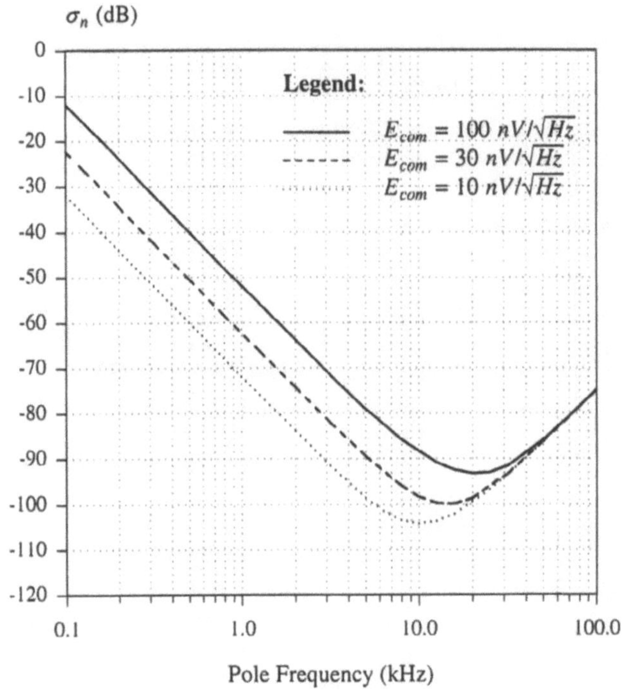

Figure 3. Baseband noise vs pole frequency with a zero at 750 kHz (2 identical poles)

remains constant. As the poles move to a higher frequency, the contribution of E_Q increases at the same rate while that of E_{com} becomes constant. Therefore, the sensitivity of the modulator's SNR to the pole frequency variation at the low or high frequency range is only 1 dB per 12 percent of pole frequency (i.e. both poles moving in the same direction). Overall, this shows that the passive sigma delta modulator is relatively insensitive to circuit parasitics.

The overall baseband noise is shown in figure 3 as a function of pole frequency. It is seen that at low frequencies E_{com} dominates the overall baseband noise while at high frequencies E_Q dominates. There exists a frequency range where both the overall baseband noise and its sensitivity to pole frequency variation are minimized. Therefore, once the comparator input referred noise E_{com} is determined, the poles of the loop filter can be chosen properly to achieve a maximum SNR.

It is also interesting to note that 2 sets of pole frequencies exist which will give the same amount of overall baseband noise. For example, with $E_{com} = 30\ nV/\sqrt{Hz}$, the modulator has a noise level at -90 dB in the baseband with poles at either 5 kHz or 40 kHz from figure 3. Therefore, a tradeoff has to be made in the design of a loop filter. If power consumption is a major concern, the loop filter poles can be placed at the higher frequency where quantization noise is dominant. In this way, E_{com} is allowed to increase further so that the

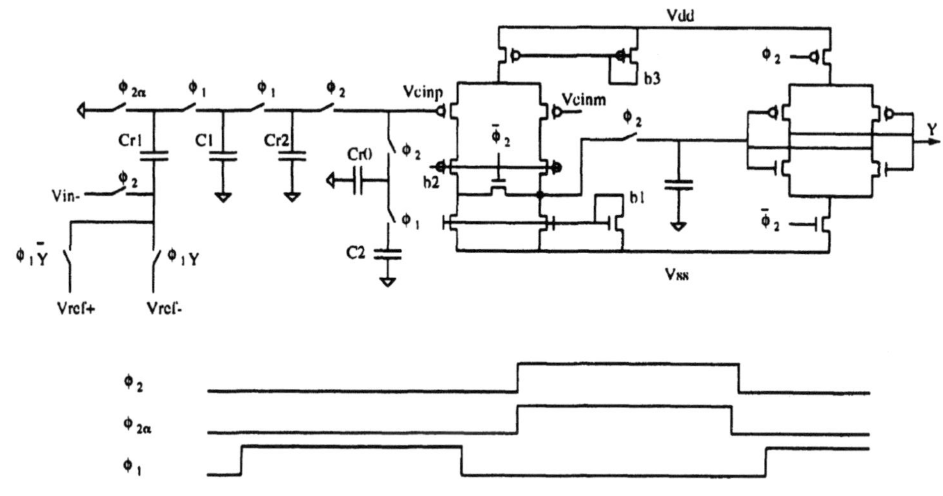

Figure 4. A SC passive sigma delta modulator and its timing diagram

power consumption of the comparator can be reduced without degrading the SNR performance. However, if resolution is a major concern, the loop filter poles can be placed at the lower frequency where E_{com} dominates thus leaving room for a higher SNR by reducing E_{com}.

3. A 2nd-Order Passive Sigma Delta Modulator

A 2nd-order passive sigma delta modulator is designed in a 1.2 um CMOS process. Only the half circuit is shown in figure 4. The loop filter is designed to have poles at 8 kHz and 34 kHz, and a zero at 750 kHz. A detailed analysis shows that this filter has a third pole at 9 MHz due to the presence of C_{com}, the equivalent input capacitance from the following comparator. Since this third pole's frequency is close to the sampling rate, it can be ignored. phi_1 and phi_2 are the two non-overlapping phases of a 10 MHz sampling clock. The clock diagram is shown in figure 4. A conventional differential architecture is adopted for better noise immunity. It consists of a preamplifier and a regenerative latch.

4. Experimental Results

The chip was tested at a 10 MHz clock rate with a 3.3 V power supply and a 5 kHz test tone. The reference voltage is 2 Vpp.

The SNDR versus the input level is shown in figure 5. A peak SNDR of 77 dB and a dynamic range of 87 dB have been measured over a bandwidth of 20 kHz (OSR=256). The measured output spectrum is shown in figure 6. It can be seen that the 2nd-order and 3rd-order harmonic distortion components are at least 90 dB down from the input fundamental which is at -11 dB. The test conditions and measured performance are summarized in table 1.

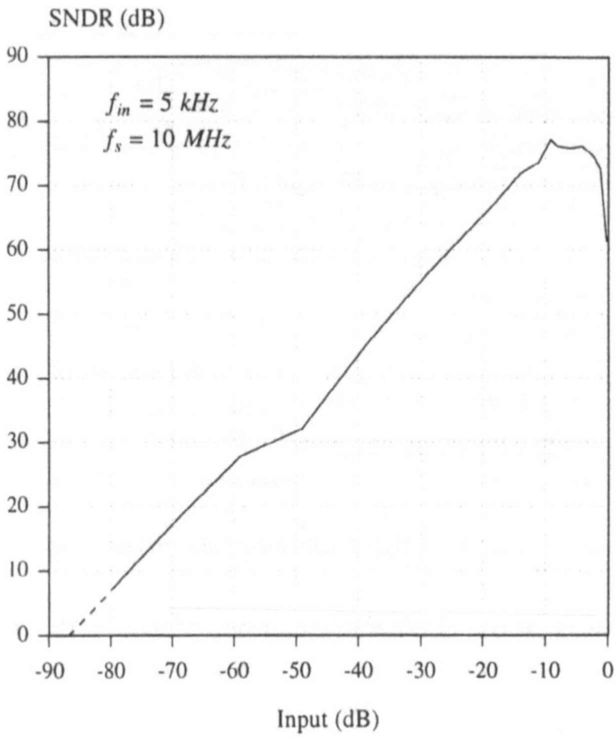

Figure 5. Measured SNDR versus the input level

Table 1. Test conditions and measured performance

IC Process	1.2 um CMOS
Active Area	$0.4\mu m$
Power Supply Voltage	3.3 V
Sampling Rate	10 MHz
Bandwidth	20 kHz
Total Bias Current	25 uA
Full Scale	2.0 V
Analog Input Freq.	5 kHz
Power Consumption	0.23 mW
Peak SNDR	77 dB
Dynamic Range	87 dB

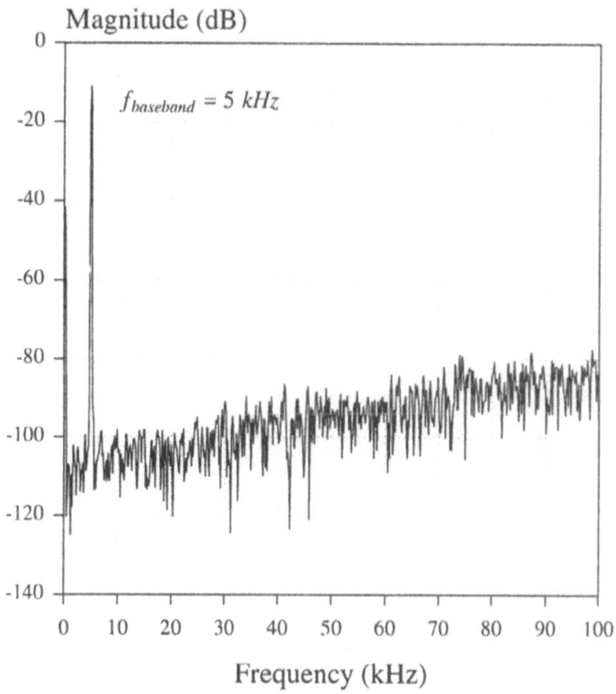

Figure 6. Measured output spectrum of the modulator with a 5 kHz input

LABELING THE NERVOUS SYSTEM WITH A BALLISTIC APPROACH

Wen-Biao Gan
Molecular Neurobiology Program
Skirball Institute
New York University School of Medicine, 540 First Avenue
New York, NY 10016

Abstract I describe here a detailed procedure for rapid labeling of a large number of cells in the nervous system. By delivering lipophilic dye coated particles to neuronal preparations with a "gene gun", individual neurons whose membranes are contacted by the particles are quickly labeled. Using particles that are each coated with different combinations of various lipophilic dyes, many cells within a complex neuronal network can be simultaneously labeled with a wide variety of colors. This ballistic technique is useful for studying structural and functional plasticity of complicated neuronal circuits.

1. Introduction

Beginning a century ago with the Golgi staining and extending into the present, neuronal labeling techniques have played a central role in neurobiology. One useful approach to label a neuron in its entirety has been to use lipophilic carbocyanine dyes [1]. These dyes are maintained within the cell membrane and can be imaged for long periods without obvious cytotoxicity [1, 2]. Although lipophilic carbocyanine dyes come in many colors, they are typically applied to the neuronal tissue in ways that label many individual cells the same color (e.g., [3, 4]). These dyes can be applied with sharp electrodes to label individual axons. However, this approach is technically tedious and only a small number of cells can be labeled at any one time [5, 6, 7].

Here, I describe a new technique that allows rapid and differential labeling of cells in various neuronal tissues by means of particle mediated ballistic delivery of lipophilic dyes such as DiO, DiI and DiD. By coating beads with various mixtures of dyes, individual neurons in living or fixed tissue can be labeled rapidly by many different colors at low or very high densities. The multicolor

The Universality of Physics: A Festschrift in Honor of Deng Feng Wang.
Edited by Khuri *et al.* Kluwer Academic/Plenum Publishers, 2001.

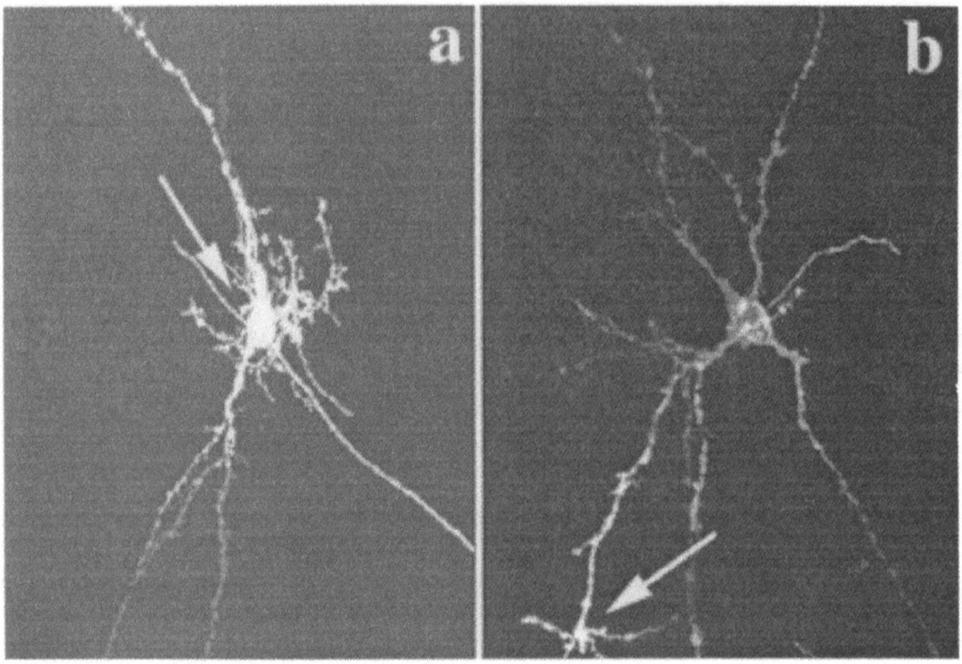

Figure 1. Labeling of cortical neurons from two different P10 mouse brain slices. a). A cell labeled by a single particle (arrow) landed on the cell soma. b). A cell labeled by a single particle landed on a dendritic arbor.

labeling facilitates the optical separation of adjacent cells. Thus this technique provides a useful means for studying interconnections of neurons.

2. Results

Tungsten or gold particles ($0.6 - 1.7\mu m$) coated with lipophilic dyes were shot into neuronal tissues with a biolistic approach similar to that used for gene transfections ([8], and see methods). In order to prevent clusters of dye-coated particles from landing on the preparation, membrane filters (see methods) with $3\mu m$ pore size and 8.0×10^5 pores/cm^2 were interposed between the "gene gun" and the target tissue. The membrane filters also appeared to protect the tissue from the shock wave generated by the gun at high pressure.

Typically we observed only one dye-coated particle juxtaposed to the cell membrane of the soma (Fig. 1a), or a branch of the dendritic or axonal arbor of a labeled cell (Fig. 1b), indicating that cell labeling does not depend on the location of the particle, in contrast to the requirement for perinuclear labeling with gene-coated particles [8]. In living tissues, neuronal dendritic trees and glial cells appeared fully labeled almost immediately (< 5 minutes) after particles contacted the tissue. Labeled dendritic arbors and axons of passage could be followed for hundreds of micrometers.

To explore the possibility of labeling neighboring neurons with different colors, we tested the appearance of neuronal preparations labeled with particles coated with various individual lipophilic dyes and dye combinations. Using three excitation wavelengths ($488nm$, $568nm$ and $647nm$ Krypton-Argon laser lines) and appropriate barrier filters ($522nm$, $590nm$ and $680nm$ respectively), cells labeled with DiI, DiO and DiD coated beads were completely distinguishable with no cross talk between the detected fluorescence with confocal laser scanning. When we bombarded tissue with particles coated with both DiO (green) and DiI (red), the labeled neurons appeared yellow or orange. Cells labeled with beads coated with DiI and DiD were detected in the red and blue channels and appeared purple. Cells that were contacted by beads coated with an equal mixture of DiO and DiD were detected in the green and blue channels and appeared cyan.

Figure 2 shows individual neurons labeled in different colors in a fixed cortical brain slice by loading the gene gun with a mixture of particles coated with 7 different combinations of dyes (see Experimental Procedures). At high magnification, neurons and their processes including dendritic spines appeared completely labeled. Although the labeling density was very high, the processes were easily distinguished from each other due to the variety of colors.

We found the labeling efficiency to be variable and dependent on multiple factors such as bullet particle density, filter pore density, gun pressure and distance from the tissue. To obtain a density of $500 - 800$ labeled cells/mm^2 in cortex, delivery of approximately 4000 particles/mm^2 was required. This was accomplished by shooting the tissue twice at 120 psi through the membrane filter. Because the tissue can be shot more than twice, the density of labeling can be increased beyond what is shown in Figure 2 (see Experimental Procedures).

3. Discussion

The technique described here allows rapid multicolor labeling of individual cells in both fixed and living neuronal preparations. It provides several advantages when compared to other labeling techniques presently in use. First, the labeling of individual cells is rapid and usually complete, allowing the entire dendritic arbor of neurons to be visualized within minutes after dye-coated particles contact the cell membrane. Axonal labeling over extensive distances takes longer but also occurs. Thus, compared to the GFP labeling techniques, which require many hours for GFP to be expressed in transfected cells [8], the ballistic technique permits immediate study of neuronal tissues as soon as the preparation is made. Such rapid labeling is especially useful in studying neuronal tissues such as brain slices that often deteriorate quickly. Furthermore,

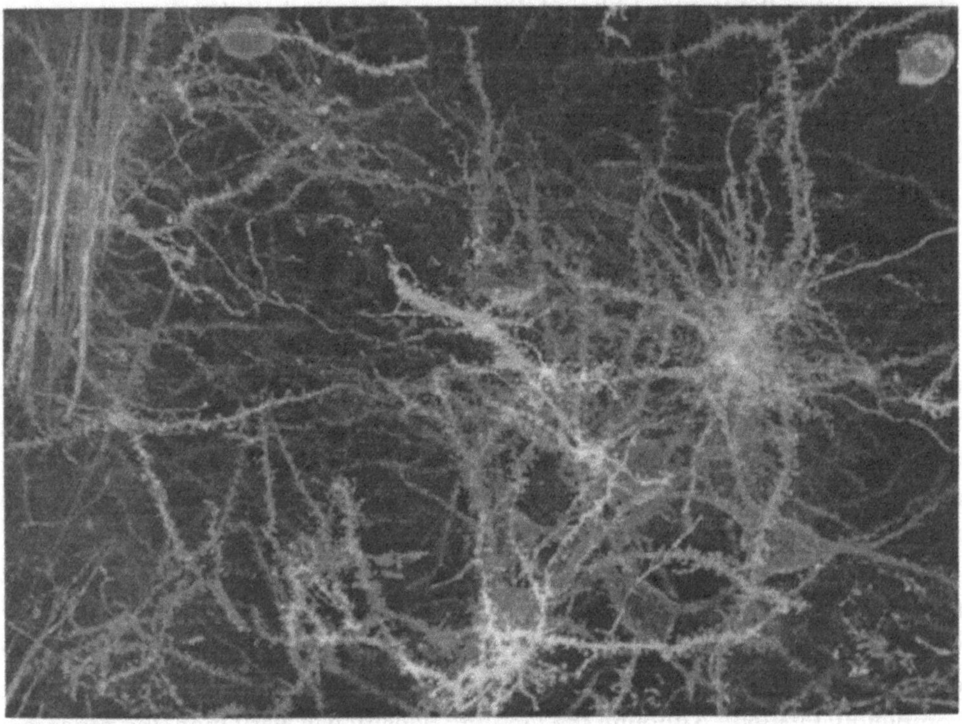

Figure 2. Image of a fixed brain slice from a P20 mouse that was shot with a combination of the seven different particle types (see Experimental Procedures). The image represents a collapsed view of many confocal planes covering approximately 50 microns of depth. Many of the cells had processes that extended beyond the upper or lower surface of the slice and hence have dendritic or axonal branches that end abruptly.

many physiological studies would benefit from an ability to rapidly label neurons and their processes prior to recording. This may be particularly useful when attempting to record from selected parts of the neuron such as from the apical or basal dendrites [9].

Second, the labeling technique involves only passive dye transfer and diffusion that is independent of gene transcription and protein synthesis. Thus, all types of neurons or glial cells can be labeled non-selectively in both fixed and living tissues, whereas the GFP labeling technique labels only those living cells which will express the GFP transgene which is under the control of a specific promotor. A further advantage of dye labeling is that individual axons of passage can also be labeled without necessity of the particles reaching the cell body as required with gene transfection techniques. Third unlike GFP, lipophilic dyes can be photoconverted relatively easily for subsequent electron microscopy [7].

Fourth, and perhaps most importantly, using various combinations of carbocyanine dyes, adjacent neurons can be labeled in different colors. Combined with high density labeling, this multicolor feature is potentially useful for studying neuronal connections in complicated neuronal networks. Because dynamic changes in processes continue to occur after dye-labeling, multicolor labeling potentially allows visualization of the interactions of pre- and postsynaptic components. In addition, neurons that appear to contact each other could be simultaneously targeted for electrophysiology in order to study their synaptic interactions.

Although the ballistic technique provides several advantages over existing vital labeling methods, it is less advantageous in certain situations. For example, neurons labeled with the carbocyanine dyes appear to bleach faster and are more susceptible to light damage compared to neurons labeled with GFP. To circumvent this problem multi-photon microscopy could be a future solution; however, imaging cells labeled with multiple dye combinations would require the selection of a set of dyes that are efficiently excited at the available infrared wavelengths. Moreover it is at present cumbersome to rapidly switch exciting wavelengths on most tunable mode-locked lasers, making the production of three-dimensional data stacks at three different exciting wavelengths difficult. Newer lasers may overcome this problem in the near future.

In summary, ballistic labeling using carbocyanine dyes provides a rapid and efficient approach to vitally label many cells in a variety of neuronal preparations. This approach should be useful in studying neuronal structure, synaptic connectivity and plasticity in the nervous system. Similar approaches may also be useful for loading cells with other indicators, such as calcium indicators, pH-sensitive dyes and other ion indicators to probe the intracellular environment.

4. Experimental Procedures

4.1. Preparations

Neonatal CF-B mice were obtained from a breeding colony (Harlan/Sprague-Dawley) maintained in our animal care facility. The date of birth was designated postnatal day 0 (P0). Mouse pups were anesthetized with $0.2ml$ sodium pentobarbital and then quickly perfused with $40ml4\%$ paraformaldehyde in 4 minutes. Brains were removed and post-fixed in 4% paraformaldehyde for 10 minutes and then sectioned with a vibratome ($200 - 300\mu m$ thick).

4.2. Preparing Particles Coated with Lipophilic Dyes:

$7mg$ DiI, DiO and DiD (Molecular Probes, Cat# D-282, D-275, D-307) are dissolved in $70\mu l$ methylene chloride (MC) in three separate $1ml$ tubes. Various volumetric proportions of different dyes are mixed to make 7 different combinations in 7 tubes as follows: (a) $30\mu l$ DiI; (b) $30\mu l$ DiO; (c) $30\mu l$ DiD; (d) $15\mu l$ DiI : $15\mu l$ DiO; (e) $15\mu l$ DiI : $15\mu l$ DiD; (f) $15\mu l$ DiO : $15\mu l$ DiD; (g) $10\mu l$ DiI : $10\mu l$ DiO : $10\mu l$ DiD. Lastly, $70\mu l$ methylene chloride is added to each tube. The final concentration of each combination is $3mg$ of dye in $100\mu l$ methylene chloride.

A small amount ($50 - 100mg$) of tungsten particles ($1.7\mu m$ diameter, Bio-Rad) is spread evenly on a glass slide by adding a few drops of methylene chloride to the particles. Each dye solution ($100\mu l$) is added to the particles on the glass slides. As methylene chloride evaporates quickly (within a minute), each dye combination precipitates onto the tungsten particles. The dye-coated particles are collected to a $10ml$ test tube using a razor blade. The particles are re-suspended in $3ml$ of distilled water and sonicated for 5-10 minutes to prevent the formation of large clusters of particles.

The " bullets" is prepared with particles of individual colors or a mixture of different color particles. This is done by injecting the sonicated solutions into plastic tubing (Bio-Rad Cat# 165-2441). It is often useful to coat the tube first with Polyvinylpyrrolidone (PVP, $10mg/ml$, from Sigma or Bio-Rad) before injecting solutions into the tube in order for better attachment of the beads to the tube. The dye-coated particles allow to precipitate and settle onto the tube wall for 5 to 15 minutes (depending on the desired density) before slowly withdrawing the remaining liquid. The particle-coated tube is then rotated for 2-5 minutes to make even distribution of beads along the tube. The tube is air-dried, cut into 13mm pieces and store at room temperature for future use.

4.3. Delivery of Particles

Dye-coated particles is delivered to the preparation using a commercially available biolistic device "gene gun" (Bio-Rad, Helios Gene Gun System Cat# 165-2431). A membrane filter with $3\mu m$ pore size and 8.0×10^5 pores/cm^2 density (Isopore membrane filter, filter type: $3.0\mu m$ TSTP, Millipore, cat no. TSTP04700) is inserted between the gene gun and the preparation to prevent clusters of large particles from landing on the tissue. The filter insertion can be made by gluing the filter to a plastic dish cover with a hole in the middle. Density of labeling is controlled by using various gas pressures (60-120 psi Helium gas, up to 200 psi) or by changing the distance between the gun and the preparation ($5 - 15mm$). Lower gas pressures and larger distance between the gun and the tissue lead to lower labeling densities. Higher gas pressure lead to breaking apart of membrane filters. Shooting the tissue several times (up to 4) is useful to increase the labeling density although it may lead to more injury when working with live tissue.

4.4. Imaging of Preparations

After delivery of particles, tissues such as brain slices are postfixed in 4% paraformaldehyde solution for 1 hour and transferred to 0.2M PBS overnight to allow dye diffusion along neuronal processes. Tissues can be mounted in 50% glycerol plus 50% 0.2M PBS or 100% glycerol on glass slides.

A Bio-Rad confocal microscope (MRC-1024) was used to image the labeled structures. A long-working distance water-immersion objective 60X, 0.9 NA or 60X, 1.4 NA oil immersion objective was used. In preparations labeled in multiple colors, we colorized as follows: DiI labeling in red, DiO in green and DiD in blue. Scanning the specimens either sequentially or simultaneously using the three excitation lines in the Kr-Argon laser (488nm for DiO, 568nm for DiI, and 647nm for DiD) with three separate barrier filter sets ($522 \pm 35nm$ for DiO, $580 \pm 32nm$ for DiI, and $680 \pm 32nm$ for DiD) gave good separation of the three image planes. A stack of images at 0.5-1.0 microns steps was acquired to generate the data set for three-dimensional neuronal structures. Each figure was generated by projecting the stack of images onto a single plane. To calibrate the colors we imaged neurons labeled with particles containing equal proportions of DiI, DiO and DiD and adjusted laser power and photodetector gain so that neurons appeared white.

Acknowledgments

Most of this work was done in Dr. Jeff Lichtman's lab. I thank Dr. Jaime Grutzendler for the collaboration in some of the experiments.

References

[1] M.G. Honig and R.I. Hume, Fluorescent carbocyanine dyes allow living neurons of identified origin to be studied in long-term cultures. J. Cell. Biol. 103, 171-187 (1986).

[2] D.W. Liu and M. Westerfield, The formation of terminal fields in the absence of competitive interactions among primary motoneurons in the zebrafish. J. Neurosci. 10, 3947-3959 (1990).

[3] N.A. O'Rourke, H.T. Cline and S.E. Fraser, Rapid remodeling of retinal arbors in the tectum with and without blockade of synaptic transmission. Neuron 12, 921-934 (1994).

[4] G.Y. Wu and H.T. Cline, Stabilization of dendritic arbor structure in vivo by CaMKII. Science 279, 222-226 (1998).

[5] W.-B. Gan and E.R. Macagno, Interactions between segmental homologues and between isoneuronal branches guide the formation of sensory terminal fields, J. Neurosci. 15, 3243-3253 (1995).

[6] W.-B. Gan and J.W. Lichtman, Synaptic segregation at the developing neuromuscular junction. Science 282, 1508-1511 (1998).

[7] W.-B. Gan, D. Bishop, S.G. Turney and J.W. Lichtman, Vital imaging and ultrastructural analysis of individual axon terminals labeled by iontophoretic application of lipophilic dye. J. Neurosci. Meth. 93(1), 13-20 (1999).

[8] D.C. Lo, A.K. McAllister and L.C. Katz, Neuronal transfection in brain slices using particle-mediated gene transfer. Neuron 13, 1263-1268 (1994).

[9] R. Yuste and D.W. Tank, Dendritic integration in mammalian neurons, a century after Cajal. Neuron 16, 701-716 (1996).

Part II

Reprints of some of Deng Feng Wang's most important papers

SPECTRUM AND THERMODYNAMICS OF THE ONE-DIMENSIONAL SUPERSYMMETRIC t-J MODEL WITH $1/r^2$ EXCHANGE AND HOPPING

D. F. Wang
Joseph Henry Laboratories of Physics
Princeton University
Princeton, New Jersey 08544

James T. Liu
Institute of Field Physics
University of North Carolina
Chapel Hill, North Carolina 27599-3255

P. Coleman
Serin Laboratories
Rutgers University
P.O. Box 849
Piscataway, New Jersey 08854

Abstract We derive the spectrum and thermodynamics of the one-dimensional supersymmetric t-J model with long-range hopping and spin exchange using a set of maximal-spin eigenstates. This spectrum confirms the recent conjecture that the asymptotic Bethe-ansatz spectrum is exact. By empirically determining the spinon degeneracies of each state, we are able to explicitly construct the free energy.

The explicit construction of low-dimensional models with Jastrow ground-state wave functions has attracted considerable recent interest [1, 2, 3, 4, 5, 6]. In one dimension, Shastry and Haldane [5, 6] have demonstrated that the ground state of the one-dimensional Heisenberg model with a $1/r^2$ exchange interaction is a Gutzwiller state for the half-filled infinite-U Hubbard model. Haldane has

Reprinted from Wang *et al.*, Phys. Rev. B 46 (1992) 6639-6642

shown how the spectrum of this model can be written in terms of a generalized type of Jastrow wave function with excitations of novel statistics [7].

Kuramoto and Yokoyama [8] have recently extended these results to include holes, demonstrating that the corresponding 1D supersymmetric t-J model is also characterized by a Gutzwiller ground state. Most recently, Kawakami has obtained an asymptotic Bethe-ansatz (ABA) solution for the model, based on the observation that the ground-state wave function is a product of two-body functions [9]. Assuming factorizability, he derived the spectrum of the system, which was conjectured to be exact. The low-energy critical behavior of the model has been identified as a Luttinger liquid [10]; the spin and charge excitations are described independently by $c = 1$ conformal field theories.

In the case of the $1/r^2$ Bose gas [11], and the Shastry-Haldane $1/r^2$ Heisenberg chain [7], the ABA has been shown to furnish the correct spectrum, despite the long-range nature of the interactions. A remarkable feature of these models is that excited states are obtained from the ground state by introducing zeros into the Jastrow wave function, in a manner reminiscent of Laughlin's description of quasiparticles in the fractional quantum Hall effect. This motivates us to examine the $1/r^2$ supersymmetric t-J model in a similar vein. Here, we show how this philosophy can be used to construct the excited-state Jastrow wave functions of the $1/r^2$ supersymmetric t-J model and indeed, the spectrum confirms Kawakami's conjecture. In addition to the spectrum, we are able to obtain the spin degeneracies of each state, permitting us to write the the free energy in closed form.

The Hamiltonian for the 1D t-J model is given by

$$H = \sum_{i\neq j,\sigma} [-t_{ij} c^\dagger_{i\sigma} c_{j\sigma}] + \sum_{i\neq j} [J_{ij}(\mathbf{S}_i \cdot \mathbf{S}_j - \tfrac{1}{4} n_i n_j)], \tag{1}$$

where we implicitly project out any double occupancies. We take $t_{ij} = J_{ij} = t/d^2(i-j)$ where $d(n) = (N/\pi)\sin(n\pi/N)$ is the chord distance consistent with periodic boundary conditions on N lattice sites [12].

States in the Hilbert space can be represented by spin and hole excitations from the fully polarized up-spin state $|P\rangle$ [13]. If we let Q denote the number of holes and M denote the number of down-spins, then S_z is given by $S_z = (N-Q)/2 - M$. The wave functions are given by

$$|\psi\rangle = \sum_{x,y} \psi(x,y) \prod_\alpha S^-_{x_\alpha} \prod_i h^\dagger_{y_i} |P\rangle, \tag{2}$$

where the amplitude $\psi(x,y)$ is symmetric in $x \equiv (x_1, x_2, \ldots, x_M)$, the positions of the down spins, and antisymmetric in $y \equiv (y_1, y_2, \ldots, y_Q)$, the positions of the holes. $S^-_{x_\alpha} = c^\dagger_{x_\alpha\downarrow} c_{x_\alpha\uparrow}$ is the spin-lowering operator at site x_α and $h^\dagger_{y_i} = c_{y_i\uparrow}$ creates a hole at site y_i.

We can construct a general class of states corresponding to states of uniform motion and spin polarization. To describe these states, we generalize Kuramoto and Yokoyama's Jastrow ground state [8] as follows:

$$\begin{aligned}\psi_G(x,y;J_s,J_h) &= \exp\left[\frac{2\pi i}{N}\left(J_s\sum_\alpha x_\alpha + J_h\sum_i y_i\right)\right]\Psi_0(x,y),\\ \Psi_0(x,y) &= \prod_{\alpha<\beta} d^2(x_\alpha - x_\beta)\prod_{i<j} d(y_i - y_j)\prod_{\alpha,i} d(x_\alpha - y_i).\end{aligned} \tag{3}$$

Here, J_s and J_h govern the (uniform) momenta of down spins and holes respectively. J_s and J_h take on either integral or half-integral values as appropriate to insure that ψ_G has the correct periodicities under $x_\alpha \to x_\alpha+N$ and $y_i \to y_i+N$.

The Hamiltonian can be broken up into four parts, $H = T^\uparrow+T^\downarrow+H^0+H^{\rm int}$, where $T^\uparrow$ ($T^\downarrow$) is the up (down) spin transfer operator, H^0 is the spin exchange operator and $H^{\rm int}$ is the diagonal interaction term. When H acts on ψ_G, $T^\uparrow$ only affects the y variables and H^0 only affects the x variables. As a result, these operators are easy to treat and yield only two- and three-body terms when appropriate conditions on J_s and J_h are met [5, 6, 8].

However, because $T^\downarrow$ exchanges pairs of x_α and y_i, this term must be treated differently. In general, it is *not* true that $T^\uparrow|\psi_G\rangle = T^\downarrow|\psi_G\rangle$ because $T^\uparrow$ does not commute with the spin raising operator. This difficulty was overlooked in earlier work [8]. To deal with $T^\downarrow$, we use an alternate representation for $|\psi_G\rangle$ in terms of up spins and holes. Let us introduce the $N-M-Q$ coordinates $u \equiv (u_1, u_2, \ldots, u_{N-M-Q})$, which give the location of the up spins. Wavefunctions in this representation are given by the spin-rotated version of Eq. (2), where the x's are replaced by u and M is replaced by $N-M-Q$. Making this transformation, we find

$$\psi_G(x,y;J_s,J_h) = A\psi_G(u,y;N-J_s,J_h-J_s+N/2), \tag{4}$$

where the set of N coordinates (x,y,u) exhausts the entire lattice. A is a constant independent of the spin and hole coordinates. Using this identity, the down-spin transfer operator gives

$$T^\downarrow\psi_G(x,y)/\psi_G(x,y) = T^\downarrow\psi_G(u,y)/\psi_G(u,y), \tag{5}$$

and can thus be treated in a similar manner as $T^\uparrow$. The result gives two- and three-body terms in the variables u and y. These terms can then be converted into sums over the x and y variables by making use of the fact that (x,y,u) runs over the entire lattice.

When the separate terms that contribute to the Hamiltonian are combined, we find that the two-body terms drop out and the three-body terms combine to give constants. As a result, ψ_G with total momentum $P = (2\pi/N)(J_s M + J_h Q)$ is an exact eigenstate of H with energy

$$\begin{aligned}(N^2/\pi^2 t)E &= \tfrac{2}{3}M(M^2-1) - 2MJ_s(N-J_s) \\ &\quad + Q[\tfrac{1}{3}(N^2-1) + \tfrac{2}{3}(Q^2-1) + \tfrac{1}{2}(M+Q)(2M-Q) \\ &\qquad -2J_h(N-J_h) + 2(J_s-J_h)^2]. \end{aligned} \tag{6}$$

The cancellation of the many-body terms, and thus this result, is only valid under the conditions $|J_s - N/2| \le N/2 - (M-1+Q/2)$, $|J_h - N/2| \le N/2 - (M+Q-1)/2$, and $|J_h - J_s| \le (M+1)/2$. For a given S_z, the minimum energy is given when J_s and J_h are as close to $N/2$ as possible. The ground state is given when S_z is either 0 or 1/2 and is a singlet whenever possible [8]. When $Q = 0$, this reduces to the result for the Heisenberg chain [5, 6]. These energy levels have also been found by Ha and Haldane [14], where $J_\uparrow = J_h - N + (M+Q+1)/2$ and $J_\downarrow = J_h - J_s - (M-1)/2$. From these energy levels we find the spin and charge velocities to be identical to the previous results [8, 14].

To investigate the other excited states of the system, we introduce zeros into the wave function by premultiplying it with polynomials of $X_\alpha = \exp(2\pi i x_\alpha/N)$ and $Y_i = \exp(2\pi i y_i/N)$. The wave functions thus take the following modified Kalmeyer-Laughlin form [15]:

$$\psi(x,y) = \mathbf{\Phi_s}(X,Y)\mathbf{\Phi_h}(Y)\mathbf{\Psi_0}, \tag{7}$$

where $\mathbf{\Phi_s}$ and $\mathbf{\Phi_h}$ are completely symmetric under pairwise interchange of their arguments. These states will be termed "fully polarized spinon states". Loosely speaking, the polynomials $\mathbf{\Phi_s}$ and $\mathbf{\Phi_h}$ can be regarded as spin and charge quasiparticle wave functions respectively.

When the Hamiltonian acts on this wave function, once again all three-body terms combine to give constants. However, in this case, some two-body terms remain and we are left with the eigenvalue equation

$$\frac{N^2}{\pi^2 t}E\mathbf{\Phi_s}\mathbf{\Phi_h} = E_0\mathbf{\Phi_s}\mathbf{\Phi_h} + H_1 + H_2 + H_3, \tag{8}$$

where

$$\begin{aligned} H_1 &= 2\mathbf{\Phi_h}\left[\sum_\mu \partial_\mu^2 + \sum_{\mu<\nu}\frac{W_\mu + W_\nu}{W_\mu - W_\nu}(\partial_\mu - \partial_\nu)\right]\mathbf{\Phi_s} \\ &\quad + 4\mathbf{\Phi_s}\left[\sum_i \partial_i^2 + \frac{1}{2}\sum_{i<j}\frac{Y_i + Y_j}{Y_i - Y_j}(\partial_i - \partial_j)\right]\mathbf{\Phi_h} \\ H_2 &= 4\sum_i \partial_i\mathbf{\Phi_s}\partial_i\mathbf{\Phi_h} \end{aligned}$$

$$H_3 = 2\mathbf{\Phi_h} \sum_{\alpha<\beta} \frac{X_\alpha + X_\beta}{X_\alpha - X_\beta}(\partial_\alpha - \partial_\beta)\mathbf{\Phi_s}, \tag{9}$$

where $W \equiv (X, Y) \equiv (X_1 \ldots X_M, Y_1 \ldots Y_{Q+M})$ and $\partial_\mu \equiv W_\mu \partial/\partial W_\mu$. In deriving this, we have shifted $\mathbf{\Phi_s}$ by the ground-state configuration, $\prod_\mu W_\mu^{N/2}$. As a result, E_0 is given by Eq. (6) with $J_s = J_h = N/2$. We require that $|\text{degree}\,\mathbf{\Phi_s}| \leq N/2-(M-1+Q/2)$, $|\text{degree}\,\mathbf{\Phi_s}\mathbf{\Phi_h}| \leq N/2-(M+Q-1)/2$, and $|\text{degree}\ \mathbf{\Phi_h}| \leq (M+1)/2$, which is to hold for each variable X_α or Y_i independently.

The first term, H_1, does not mix $\mathbf{\Phi_s}$ and $\mathbf{\Phi_h}$ and has been solved by Sutherland [11]. However, H_2 mixes $\mathbf{\Phi_s}$ and $\mathbf{\Phi_h}$ and H_3 does not act symmetrically on $\mathbf{\Phi_s}$. As a result, they are harder to deal with. We follow Sutherland and start by choosing the following symmetric basis functions:

$$\begin{aligned}\mathbf{\Phi_s}(W;\{n\}) &= \sum_{\{P_\mu\}} \prod_\mu W_\mu^{n_{P_\mu}} \\ \mathbf{\Phi_h}(Y;\{m\}) &= \sum_{\{P_i\}} \prod_i Y_i^{m_{P_i}},\end{aligned} \tag{10}$$

where the quantum numbers $\{n_1, \ldots, n_{M+Q}\}$ and $\{m_1, \ldots, m_Q\}$ are taken to be in increasing order and $\{P_\mu\}$ and $\{P_i\}$ denote permutations of the indices. These quantum numbers are integral or half-integral as required by periodic boundary conditions.

In this basis, labeled by the two sets of quantum numbers $\{n_\mu\}$ and $\{m_i\}$, the Hamiltonian, considered as a matrix, can be shown to be upper triangular. Eigenvalues are found by reading the diagonal elements labeled in terms of the quantum numbers $\{n_\mu\}$ and $\{m_i\}$ [16]. The result simplifies when written in terms of a conjugate set of quantum numbers $\{J_1, J_2, \ldots, J_{M+Q}\}$ and $\{I_1, I_2, \ldots, I_Q\}$ defined by

$$\begin{aligned} J_\mu &= n_\mu + n_\mu^0, \qquad & n_\mu^0 &= \frac{1}{2}(2\mu - (M+Q) - 1) \\ I_i &= m_i + m_i^0, \qquad & m_i^0 &= \frac{1}{2}(2i - Q - 1),\end{aligned} \tag{11}$$

where $\{n_\mu\}$ and $\{m_i\}$ must satisfy the conditions specified before. This translates into the conditions $|J_\mu| \leq (N - M + 1)/2$ and $|I_i| \leq (M+Q)/2$. The energy is

$$\frac{E}{t} = \frac{\pi^2}{3}Q\left(1 - \frac{1}{N^2}\right) + \frac{1}{2}\sum_{\mu=1}^{M+Q}(p_\mu^2 - \pi^2), \tag{12}$$

where the pseudomomenta p_μ are given by the following equations:

$$\begin{aligned} p_\mu N &= 2\pi J_\mu - \pi \sum_{i=1}^{Q} \mathrm{sgn}(p_\mu - q_i) + \pi \sum_{\nu=1}^{M+Q} \mathrm{sgn}(p_\mu - p_\nu), \\ 2\pi I_i &= \pi \sum_{\mu=1}^{M+Q} \mathrm{sgn}(q_i - p_\mu). \end{aligned} \tag{13}$$

The above equations correspond to the asymptotic Bethe-ansatz equations obtained by Kawakami [9]. Our result thus confirms that the ABA spectrum is exact.

Here the resulting $\{p_\mu\}$ and $\{q_i\}$ must lie between $-\pi$ and π. The set of $M+Q$ distinct quantum numbers J_μ is in ascending order and governs the spin excitations. We restrict them to take values in the range $[-(N-M-1)/2, (N-M-1)/2]$ to guarantee that they generate fully spin-polarized states. There are $N-M$ values in this range, of which $M+Q$ are occupied and $2S_z$ are empty. A spin configuration can be represented by a sequence of $N-M$ digits such as $\{S\} = (0111001011)_s$, where 1 represents an occupied quantum number and 0 an unoccupied quantum number. These empty values are identified as spinons [7]; a sequence of $2j_r$ consecutive zeros corresponds to a symmetric bound-state of $2j_r$ spinons, thereby creating an excitation of spin j_r with spin degeneracy $2j_r + 1$. On these physical grounds, we anticipate a spin degeneracy in the thermodynamic limit given by

$$w_s = \prod_j (2j+1)^{n(j)}, \tag{14}$$

where $n(j)$ is the number of sequences of zeros of length $2j$. The set of Q distinct quantum numbers I_i in ascending order and taking values in the range $[-(M+Q)/2, (M+Q)/2]$, governs charge excitations.

To complete the study of the model and confirm our interpretation of the quasiparticle degeneracies, we looked at exact diagonalization of small systems ($N \leq 10$ with holes). As an example, the low-lying states of the $N = 10, Q = 2$ model is shown in Fig. 1. We summarize the numerical result as follows:

1 The spectrum described in terms of the real pseudomomenta $\{p_\mu\}$ and $\{q_i\}$ span the *full* set of energy levels of the system.

2 The real pseudomomentum states are all highest weight states when $\{p_\mu\} \neq \pm\pi$.

3 The spin degeneracy rule is obeyed for all internal sequences of zeros.

Certain small corrections to the spin degeneracy rule apply when there are zeros at either end of $\{S\}$ which we shall not enumerate here, and which are not important in the thermodynamic limit [16].

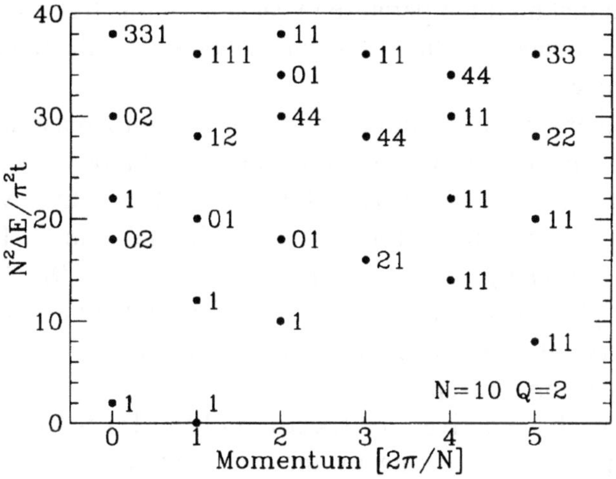

Figure 1. Low-lying energy levels of the 10 site 2 hole system from exact diagonalization. The numbers associated with each state list the spin degeneracies starting with spin 0 on the left. For example, the number "331" indicates that we have 3 states with $S = 0$, 3 states with $S = 1$ and 1 state with $S = 2$.

Finally, we may use the spectrum generated by the "fully polarized spinon states" and the supermultiplicity rule to obtain the free energy of the model in the thermodynamic limit. Besides the "particle-state" solutions of Eqs. (12) and (13), we have to take into account the "hole-state" solutions [17]. At thermal equilibrium, the distribution functions of these solutions are determined by minimizing the free-energy functional [18, 19], $F = E - TS - \mu(N - Q)$, with the constraint that each "fully polarized spinon state" described by quantum numbers $\{J_\mu\}$ and $\{I_i\}$ is associated with a spin degeneracy w_s as given in (14), where μ is the chemical potential. Following the standard methods of Takahashi [18], minimizing the free energy for a given quantum number distribution gives the following free energy:

$$F(T)/N = -\mu - \frac{T}{2\pi}\int_{-\pi}^{\pi} dp\, \ln(1 + e^{-\beta\epsilon_s(p)}), \tag{15}$$

where ϵ_s is determined by the coupled equations

$$\begin{aligned} 2\epsilon_s(p) &= \epsilon_0(p) - 2a - T\ln(1 + e^{-\beta\epsilon_c(p)}) \\ \epsilon_c(q) &= 2a - T\ln(1 + e^{-\beta\epsilon_s(q)}). \end{aligned} \tag{16}$$

Here $\epsilon_0(p) = \frac{1}{2}t(p^2 - \pi^2/3) + \mu$ and $a = \frac{1}{6}t\pi^2 + \frac{1}{2}\mu$. In the limit of half filling, $\mu \to \infty$, $\epsilon_s \to t(p^2 - \pi^2)/4$, and the free energy reverts to the form obtained by Haldane for the corresponding Heisenberg model [7, 20]. For general μ, elimination of $\epsilon_c(p)$ yields the result

$$\epsilon_s(p) = \epsilon_o(p) - T\ln(\tfrac{1}{2} + [\tfrac{1}{4} + 2e^{\beta[\epsilon_0(p)+a]}\cosh(\beta a)]^{1/2}), \tag{17}$$

We have verified that high-temperature expansion of this free energy in powers of β correctly reproduces the first two nontrivial terms in the high-temperature perturbation theory.

From our free energy, it is not clear whether we may make a unique identification of the statistics of the spin and charge excitations. We note that the $S = 1/2$ spinon excitations always combine into a state with a symmetric spin wave function, thus $2S$ spinons form a state with total spin S. In the limit of zero doping, the free energy is that of spinless fermions [7, 20]. We can equally well regard the spinon excitations as $S = 1/2$ fermions in a state with a fully antisymmetric spatial wave function, or alternatively, as hardcore $S = 1/2$ bosons in a fully symmetric spatial state.

In summary, we have derived the spectrum of the 1D t-J model with $1/r^2$ long-range exchange and hopping by the introduction of zeros into Jastrow ground-state wave functions. Our solution confirms Kawakami's conjecture that the ABA provides the exact spectrum, suggesting that despite the long-range nature of the interactions, two-body scattering dominates the long-wavelength physics. By interpreting multiple occupancy of momentum states in the spinon wave function as symmetric bound complexes of spinons, we have been able to determine the degeneracies of the states needed to construct the free energy. Further work is required to determine the integrability conditions of this model. There are also several possible generalizations: most notably, SU(N) generalizations and the appealing possibility of Jastrow-integrable impurity models.

Acknowledgments

We would like to thank Z. Ha and F. D. M. Haldane for informative discussions. This work was supported in part by the U.S. Department of Energy under Grant No. DE-FG05-85ER-40219, the National Science Foundation under Grant NSF-DMR-13692, and by a Sloan Foundation Grant.

References

[1] R. B. Laughlin, Phys. Rev. Lett. **50** ,1395 (1983).

[2] D. Arovas and S. M. Girvin (unpublished).

[3] A. Auerbach, D. Arovas and D. Haldane, Phys. Rev. Lett. **60**, 531 (1988).

[4] C. S. Hellberg and E. J. Mele, Phys. Rev. B **44**, 1360 (1991); Phys. Rev. Lett. **67**, 2080 (1991); *Lecture Notes in Physics*, edited by B. S. Shastry *et al.*, (Spring-Verlag, New York, 1985) Vol. 242, p. 1.

[5] F. D. M. Haldane, Phys. Rev. Lett. **60**, 635 (1988).

[6] B. S. Shastry, Phys. Rev. Lett. **60**, 639 (1988).

[7] F. D. M. Haldane, Phys. Rev. Lett. **66**, 1529 (1991).

[8] Y. Kuramoto and H. Yokoyama, Phys. Rev. Lett. **67**, 1338 (1991).

[9] N. Kawakami, Phys. Rev. B **45**, 7525 (1992).

[10] N. Kawakami and S. K. Yang, Phys. Rev. Lett. **67**, 2493 (1991).

[11] B. Sutherland, Phys. Rev. A **5**, 1372 (1972); **4**, 2019 (1971); J. Math. Phys. **12**, 251 (1971); **12**, 246 (1971).

[12] Note that a similar model with $1/r$ long-range hopping and a finite local repulsive U has recently been considered by F. Gebhard and A. E. Ruckenstein, Phys. Rev. Lett. **68**, 244 (1992).

[13] P. W. Anderson, B. S. Shastry and D. Hristopulos, Phys. Rev. B **40**, 8939 (1989).

[14] Z. Ha and F. D. M. Haldane (unpublished).

[15] V. Kalmeyer and R. B. Laughlin, Phys. Rev. Lett. **59**, 2095 (1987).

[16] J. T. Liu, (unpublished).

[17] C. N. Yang and C. P. Yang, J. Math. Phys. **10**, 1115 (1969).

[18] M. Takahashi, Prog. Theor. Phys. **46**, 401 (1971).

[19] H. M. Babudjan, Nucl. Phys. **B215** [FS7], 317 (1983).

[20] F. D. M. Haldane and A. Tsvelik (unpublished); F. Gebhard and A. E. Ruckenstein, Phys. Rev. Lett. **68**, 244 (1992).

GUTZWILLER-JASTROW WAVE FUNCTIONS FOR THE $1/r$ HUBBARD MODEL

D. F. Wang

Joseph Henry Laboratories of Physics
Princeton University
Princeton, New Jersey 08544

Q. F. Zhong

International School for Advanced Study
Via Beirut 4
34014 Trieste
Italy

P. Coleman

Serin Physics Laboratory
Rutgers University
P. O. Box 849
Piscataway, New Jersey 08854

Abstract In this work, we study the wave functions of the one-dimensional $1/r$ Hubbard model in the strong-interaction limit $U = \infty$. A set of Gutzwiller-Jastorw wave functions are shown to be eigenfunctions of the Hamiltonian. The entire excitation spectrum and the thermodynamics are also studied in terms of more generalized Jastrow wave functions. For the wave functions and integrability conditions at finite on-site energy, further investigations are needed.

There has been considerable interest in the study of low-dimensional integrable models [1, 2, 3, 4, 5, 6, 7, 8, 9, 10, 11, 12, 13, 14, 15, 16]. One is the Hubbard model with $1/r$ hopping and U on-site energy, introduced by Gebhard and Ruckenstein [9]. In this work we report some new results in our recent study of the system.

The long-range Hubbard model is exactly solvable in one dimension for arbitrary on-site energy [9]. Based on small-size numerical work and several limiting cases, Gebhard and Ruckenstein have introduced an effective Hamilto-

Reprinted from Wang *et al.*, Phys. Rev. B 48 (1993) 8476-8479

nian, which was conjectured to be equivalent in describing the Hubbard model for arbitrary on-site energy. With the effective Hamiltonian, the system has been found to exhibit a metal-insulator transition when the bandwidth $2\pi t$ is equal to the on-site energy U in the half-filling case, and the Gutzwiller state is an eigenstate of the effective Hamiltonian in the large-U limit. The complete wave functions and the integrability conditions have remained unsolved for a long time.

In this paper, we only concentrate on the study of the wave functions and the spectrum in the strong interaction limit $U = \infty$. In this extreme limit, one unfortunate artifact is that the spins are completely free and decoupled trivially from the charge degrees of freedom. We introduce a set of generalized Gutzwiller-Jastrow wave functions, and we show that they are exact eigenfunctions of the Hamiltonian. The lowest energy state in this set is also a ground state of the system in the whole Hilbert space. Considering more generalized Jastrow wave functions, we are able to write the full spectrum. The separation of spin and charge in the full excitation spectrum shows that the system is a special example of Luttinger liquids in the sense of Haldane.

The Hamiltonian for the one-dimensional Hubbard model is given by

$$H = \sum_{i \neq j; \sigma = \uparrow,\downarrow} t_{ij} c_{i\sigma}^{\dagger} c_{j\sigma} + U \sum_{i} n_{i\uparrow} n_{i\downarrow}, \tag{1}$$

where $c_{i\sigma}^{\dagger}$ and $c_{i\sigma}$ are creation and annihilation operators at site i with spin component σ. We take $t_{ij} = it(-1)^{(i-j)}/d(i-j)$ where $d(n) = (L/\pi)\sin(n\pi/L)$ is the chord distance [9]. Here we assume periodic boundary condition for the wave functions for odd L, or antiperiodic boundary condition for even L.

In the strong-interaction limit $U = \infty$, each site can be occupied by at most one electron. In this limit, the Hamiltonian can be written in terms of the Hubbard operator as

$$H = \sum_{i \neq j; \sigma = \uparrow,\downarrow} t_{ij} X_i^{\sigma 0} X_j^{0\sigma}. \tag{2}$$

In the following, whenever in the case of the strong interaction, we always implicitly assume no double occupancy. Let us denote the number of holes by Q, the number of down spins by M. Following notations used previously, the state vectors can be represented by creating spin and charge excitations from the fully polarized up-spin state $|P\rangle$ [10],

$$|\Phi\rangle = \sum_{\alpha, j} \Phi(\{x_\alpha\}, \{y_j\}) \prod_{\alpha} b_{\alpha}^{\dagger} \prod_{j} h_j^{\dagger} |P\rangle, \tag{3}$$

where $b_{\alpha}^{\dagger} = c_{\alpha\downarrow}^{\dagger} c_{\alpha\uparrow}$ is the operator to create a down spin at site α, and $h_j^{\dagger} = c_{j\uparrow}$ creates a hole at site j. The amplitude $\Phi(\{x_\alpha\}, \{y_j\})$ is symmetric in the down-spin positions, and antisymmetric in hole positions.

To describe uniform motion and magnetization, we consider the following generalized Gutzwiller-Jastrow wave functions:

$$\Phi(x, y; J_s, J_h) = \left\{ \exp\left[\frac{2\pi i}{L}\left(J_s \sum_\alpha x_\alpha + J_h \sum_i y_i\right)\right] \Phi_0 \right\},$$
$$\Phi_0 = \prod_{\alpha<\beta} d^2(x_\alpha - x_\beta) \prod_{\alpha i} d(x_\alpha - y_i) \prod_{i<j} d(y_i - y_j), \tag{4}$$

where the quantum numbers J_s and J_h govern the momenta of the down spins and holes, respectively. They can be integers or half-integers so that we have appropriate periodicities (or antiperiodicities) for the wave functions under the translations $x_\alpha \to x_\alpha + L$, or $y_i \to y_i + L$ for odd L (or even L). For the wave functions to be exact eigenfunctions of the Hamiltonian, the momenta of down spins and holes must take values from some restricted regions, which will be specified below.

These Gutzwiller-Jastrow wave functions have been studied extensively in the recent study of the integrable models of long-range interaction [6, 7, 10, 11, 12]. They were found to be the exact eigenfunctions of those systems. In the following, we demonstrate that they are also the exact solutions of the Hubbard model. To show that they are eigenfunctions of the Hamiltonian, we have to treat the hopping operator very carefully.

The hopping operator can be broken up into two parts, the up-spin hopping operator and the down-spin hopping operator. For the up-spin hopping operator $T_\uparrow = \sum_{i\neq j} t_{ij} c_{i\uparrow}^\dagger c_{j\uparrow}$, its effect is just the hopping of holes alone when it operates on the wave functions. But the down-spin hopping operator, $T_\downarrow = \sum_{i\neq j} t_{ij} c_{i\downarrow}^\dagger c_{j\downarrow}$, will involve the hopping of down-spins and holes simultaneously, which needs to be treated using the spin-rotated version of the Gutzwiller-Jastrow wave functions, developed in the recent work on the supersymmetric t-J model [12].

For the up-spin hopping, we have

$$\frac{T_\uparrow \Phi(x, y; J_s, J_h)}{\Phi(x, y; J_s, J_h)} = -it \sum_i \sum_{n=1}^{L-1} \frac{(-1)^n}{d(n)} z^{nJ_h} \prod_{j(\neq i)} F_{ij}(n) \prod_\alpha F_{i\alpha}(n), \tag{5}$$

where $z = \exp(2\pi i/L)$, $F_{ij}(n) = \cos(\pi n/L) + \sin(\pi n/L) \cot\theta_{ij}$, $F_{i\alpha}(n) = \cos(\pi n/L) + \sin(\pi n/L) \cot\theta_{i\alpha}$, with $\theta_{ij} = \pi(y_i - y_j)/L$, $\theta_{i\alpha} = \pi(y_i - x_\alpha)/L$. The sum can be calculated in a standard way, by expanding the products and classifying the terms by the number of particles involved. The terms with more than two particles vanish, yielding the following result:

$$\frac{T_\uparrow \Phi(x, y; J_s, J_h)}{\Phi(x, y; J_s, J_h)} = -(2\pi t/L) J_h Q + (2\pi t/L) i \sum_{\alpha,i} \cot\theta_{i\alpha} \tag{6}$$

where the hole momenta satisfy the condition $|J_h| \le L/2 - (M+Q)/2$.

To deal with the down-spin hopping operators, using the spin-rotated version of the Gutzwiller-Jastrow wave functions, we find

$$\frac{T_{\downarrow}\Phi(x,y;J_s,J_h)}{\Phi(x,y;J_s,J_h)} = -it\sum_i \sum_{n=1}^{L-1} \frac{(-1)^n}{d(n)} z^{n\tilde{J}_h} \prod_{j(\neq i)} F_{ij}(n) \prod_{\mu} F_{i\mu}(n), \quad (7)$$

where $F_{i\mu}(n) = \cos(\pi n/L) + \sin(\pi n/L)\cot\theta_{i\mu}$, $\tilde{J}_h = J_h - J_s + L/2$, $\theta_{i\mu} = \pi(y_i - u_\mu)/L$, for $\mu = 1, 2, \ldots, L-M-Q$. Here $u_1, u_2, \ldots, u_{L-M-Q}$ are the positions of the up spins on the lattice. Finally, we obtain

$$\frac{T_{\downarrow}\Phi(x,y;J_h,J_s)}{\Phi(x,y;J_h,J_s)} = -(2\pi t/L)\tilde{J}_h Q + (2\pi t/L)i\sum_{\mu,i}\cot\theta_{i\mu}. \quad (8)$$

Here $|\tilde{J}_h| \le L/2 - (\tilde{M}+Q)/2$, $\tilde{M} = L - Q - M$.

After adding the spin-up and spin-down hopping operator effects together, the two-particle terms vanish due to the fact that the down-spin electrons, the up-spin electrons and the holes span the entire lattice. As a result, we see that the Gutzwiller-Jastrow wave functions are eigenstates of the Hamiltonian, with eigenenergies given by

$$E(J_s, J_h) = -(2\pi t/L)[2J_h - J_s + L/2]Q. \quad (9)$$

If we take t to be positive, the lowest energy in this set of Jastrow wave functions is obtained when J_h and $\tilde{J}_h$ reach their allowed maximum possible values, which occurs at $J_h = L/2 - (M+Q)/2$ and $J_s = L - M - Q/2$. Thus the dependence of ground state energy on the number of holes Q is given by

$$E_0 = -(2\pi t/L)[L/2 - Q/2]Q. \quad (10)$$

This energy is also the absolute ground state energy of the system in the whole Hilbert space [9], as indicated by our numerical results of six- and eight-site lattices. For this ground-state wave function, the exponents of the long-distance behaviors of various correlators can be obtained from the recent work by Kawakami. In general, the ground state is not unique, and this Jastrow wave function ground state is only one of them. The ground state degeneracy is given by $(M+\tilde{M})!/(M!\tilde{M}!)$.

As seen above, the energy consists of decoupled down-spin contribution and the hole contribution. In principle, we may find more generalized Jastrow wave functions as eigenfunctions of the Hamiltonian. As will be seen below, in our case, we may even write the entire spectrum of the system in the full Hilbert space in terms of a set of Jastrow functions.

To study other excitations, let us consider more generalized Jastrow wave functions in the form

$$\phi = \phi_s(X,Y)\phi_h(Y)\Phi_0. \quad (11)$$

Here the functions ϕ_s and ϕ_h are polynomials of $X = \{\exp(2\pi x_\alpha i/L)\}$, $Y = \{\exp(2\pi y_i i/L)\}$. They are totally symmetric in their arguments, respectively. The degrees of these polynomials must satisfy some specific conditions, so that many-particle terms vanish when applying the hopping operators. The eigenenergy equation thus reduces to

$$-(2\pi t/L)\left[\sum_{i=1}^{Q} \partial_i(\phi_s\phi_h) + \sum_{i=1}^{Q} \phi_s(\partial_i + L/2)\phi_h\right] = E\phi_s\phi_h, \qquad (12)$$

where $\partial_i = Y_i\partial/\partial Y_i$. This eigenvalue equation can be solved exactly, yielding the spectrum given by

$$E = -(2\pi t/L)\left[\sum_{i=1}^{Q} n_i + \sum_{\mu=1}^{Q} m_\mu\right]. \qquad (13)$$

Here, the integers (or half-integers) satisfy the conditions $|n_i| \leq L/2 - (\tilde{M} + Q)/2$, $|m_\mu| \leq L/2 - (M+Q)/2$, where $n_i \leq n_{i+1}, m_\mu \leq m_{\mu+1}$. This result shows that the spectrum is invariant when changing the sign of t.

We may write the spectrum in terms of a set of conjugate quantum numbers defined by $K_i = n_i + m_i + (L-Q)/2 + (i)$. The spectrum is given by

$$E = -(2\pi t/L)\sum_{i=1}^{Q} K_i + (\pi t Q/L)(L+1), \qquad (14)$$

where K_i takes values from $(1, 2, \ldots, L)$. They may be regarded as the momenta of quasiparticle "holons" [9]. Our numerical results of six- and eight-site lattices indicate that the above spectrum spans the full spectrum in the entire Hilbert space.

In the spectrum formula, each energy level is determined by a charge configuration, such as (101010) for $L = 6$, $Q = 3$, where the 1's represent the charge momenta. We may regard the empty values as the momenta for the quasiparticles of spin degrees. Our numerical result shows that the degeneracy of each energy level is given by the number of the ways to distribute $L - Q - M$ spins $s = +\frac{1}{2}$ and M spins $s = -\frac{1}{2}$ among the empty values. Thus we see that the spin degree is decoupled from the charge degree of freedom for the entire excitation spectrum [9].

For fixed number of electrons $N_e = L - Q$ on the lattice, the free energy consists of two parts, $F = F_1 - TN_e \ln 2$, where the second term comes from the decoupled spin degrees of freedom, and F_1 is the contribution from the charge degree of freedom, which is that of Q spinless fermions with the relativistic spectrum. In the grand canonical case [17, 18, 19], denoting the chemical potential of the electrons by μ, we find that the free energy per lattice site is given by

$$F(T,\mu)/L = -\mu - \frac{T}{2\pi}\int_{-\pi}^{\pi} dq \, \ln[2 + e^{\beta(qt-\mu)}]. \tag{15}$$

Here the number of electrons as a function of the chemical potential is given by $N_e = -\partial F(T,\mu)/\partial\mu$. We have also checked that the free energy correctly reproduces the first three terms in the high-temperature perturbation expansion.

We have seen that the spin degrees of freedom decouple from the charge degrees of freedom for this system in the full excitations in the strong-interaction limit. One similar system has been studied by one of the authors. He has introduced another integrable model, the multi-component Hubbard model of $1/r$ hopping and U on-site energy, which is the direct generalization of the model to the SU(N) case, where a set of generalized SU(N) Gutzwiller-Jastrow wave functions has been shown to be exact eigenfunctions of the Hamiltonian in the strong-interaction limit [13].

In summary, we have studied the wave functions of the one-dimensional $1/r$ Hubbard model in the limit of strong interaction $U = \infty$. We have shown that a set of Gutzwiller-Jastrow wave functions are the eigenstates of the Hamiltonian, and the lowest energy in this set is the ground-state energy in the whole Hilbert space. The full spectrum can be written in terms of more generalized Jastrow wave functions. The spin-charge separation occurs in the full excitation spectrum, and the system may be regarded as a special Luttinger liquid.

Finally, we would like to stress that the model is completely integrable for arbitrary on-site energy U [9]. The most interesting thing is to investigate the wave functions and the integrability conditions for this $1/r$ Hubbard model at finite on-site energy U. It would be also of great interest to study the ground-state properties, such as the spin and charge susceptibilities, and various ground-state correlation exponents dependent on the on-site energy U. For these finite on-site energy studies, further investigations are needed.

Acknowledgments

This work was supported in part by the National Science Foundation under Grant No. NSF-DMR-89-13692.

References

[1] D. Arovas and S. M. Girvin (unpublished); A. Auerbach, D. Arovas, and D. Haldane, Phys. Rev. Lett. **60**, 531 (1988); C. S. Hellberg and E. J. Mele, Phys. Rev. B **44**, 1360 (1991); Phys. Rev. Lett. **67**, 2080 (1991); in *Exactly Solvable Models*, edited by B. S. Shastry *et al.*, Lecture Notes in Physics Vol. 242 (Spring-Verlag, Berlin, 1985), p. 1.

[2] F. D. M. Haldane, Phys. Rev. Lett. **60**, 635 (1988); **66**, 1529 (1991).

[3] S. Shastry, Phys. Rev. Lett. **60**, 639 (1988).

[4] A. P. Polychronakos, Phys. Rev. Lett. **69**, 703 (1992); **70**, 2329 (1993).

[5] M. Fowler and J. A. Minahan, Phys. Rev. Lett. **70**, 2325 (1992).

[6] Y. Kuramoto and H. Yokoyama, Phys. Rev. Lett. **67**, 1338 (1991).

[7] N. Kawakami, Phys. Rev. B **47**, 2928 (1993); N. Kawakami and S. K. Yang, Phys. Rev. Lett. **67**, 2493 (1991); N. Kawakami, Phys. Rev. B **45**, 7525 (1992); **46**, 3191 (1992); **46**, 1005 (1992).

[8] B. Sutherland, Phys. Rev. A **5**, 1372 (1972); **4**, 2019 (1971); J. Math. Phys. **12**, 251 (1971); **12**, 246 (1971).

[9] F. Gebhard and A. E. Ruckenstein, Phys. Rev. Lett. **68**, 244 (1992).

[10] P. W. Anderson, B. S. Shastry, and D. Hristopulos, Phys. Rev. B **40**, 8939 (1989).

[11] Z. N. C. Ha and F. D. M. Haldane, Phys. Rev. B **46**, 9359 (1992).

[12] D. F. Wang, James T. Liu, and P. Coleman, Phys. Rev. **B 46**, 4663 (1992).

[13] D. F. Wang (unpublished).

[14] Bill Sutherland and B. Sriram Shastry (unpublished).

[15] B. Sriram Shastry and B. Sutherland (unpublished).

[16] Holger Frahm (unpublished).

[17] C. N. Yang and C. P. Yang, J. Math. Phys. **10**, 1115 (1969).

[18] M. Takahashi, Prog. Theor. Phys. **46**, 401 (1971).

[19] H. M. Babudjan, Nucl. Phys. **B215** [FS7], 317 (1983).

SOLUTIONS TO THE MULTIPLE-COMPONENT $1/r$ HUBBARD MODEL

D. F. Wang
Joseph Henry Laboratories of Physics
Princeton University
Princeton, New Jersey 08544

Abstract In this work we introduce a one-dimensional multiple-component Hubbard model with $1/r$ hopping and on-site energy U. The wave functions, the spectrum, and the thermodynamics are studied for this model in the strong-interaction limit $U \to \infty$. In this limit, the system is a special example of SU(N) Luttinger liquids, exhibiting spin-charge separation in the full Hilbert space. Speculations on the physical properties of the model at finite on-site energy are also discussed.

Recent studies on low-dimensional systems have renewed great interest in the Gutzwiller-Jastrow wave functions. The wave functions are useful in the sense that they may serve as good variational wave functions or they may be exact solutions of the Hamiltonians [1, 2, 3, 4, 5, 6, 7, 8, 9, 10, 11, 12, 13, 14, 15, 16, 17, 18, 19]. Most recently, intense research in the field of integrable systems has shown the wave functions to be exact solutions of some quantum many-particle systems. These integrable systems are characterized by the fact that the full spectrum may even be written in terms of more generalized Jastrow wave functions, as in the cases of $1/r^2$ Fermi or Bose gases, $1/r^2$ Haldane-Shastry spin chain, and the 1D supersymmetric t-J model with $1/r^2$ hopping and exchange.

The Hubbard model has been of great interest since the discovery of high-T_c superconductivity. About two years ago, Gebhard and Ruckenstein introduced the one-dimensional SU(2) Hubbard model with $1/r$ hopping and on-site energy U (Ref. [11]). The model is completely integrable for arbitrary on-site energy. In the strong-interaction limit $U \to \infty$, it has been discovered recently that a set of Gutzwiller-Jastrow wave functions is an exact set of eigenfunctions of the Hamiltonian [16], and that the system exhibits spin-charge separation in the full Hilbert space [11, 16].

Reprinted from Wang, Phys. Rev. B 48 (1993) 10556-10558

In this work, we introduce an integrable model, the one-dimensional $1/r$ multiple-component Hubbard model. In the following we only discuss the strong-interaction limit $U \to \infty$. Generalizing our previous work, we show that a set of SU(N) Gutzwiller-Jastrow wave functions is a set of eigenstates of the system. The full excitation spectrum and the thermodynamics are also given explicitly in this strong-interaction case. Spin and charge are decoupled in the full Hilbert space and the system is a special example of SU(N) Luttinger liquids. At the end of the work, we also discuss speculations of further investigation of the system of finite on-site energy U.

The Hamiltonian for the one-dimensional Hubbard model is given by

$$H = \sum_{\sigma} \sum_{i \neq j} t_{ij} c_{i\sigma}^{\dagger} c_{j\sigma} + U \sum_{i} \sum_{\sigma \neq \sigma'} n_{i\sigma} n_{i\sigma'}, \tag{1}$$

where $c_{i\sigma}^{\dagger}$ and $c_{i\sigma}$ are creation and annihilation operators at site i with spin component σ. The sum over σ runs from 1 to N, where N is the number of flavors of the fermions. We take the hopping matrix $t_{ij} = it(-1)^{(i-j)}/d(i-j)$ where $d(n) = (L/\pi)\sin(n\pi/L)$, and U is the on-site energy. Here, because of the special form of the hopping matrix for the wave functions of the system, we assume periodic boundary conditions for odd L, or antiperiodic boundary condition for even L.

In the strong-interaction limit $U \to \infty$, each site can be occupied at most by one particle. In this limit, we work in the Hilbert space of no double occupancy and no multiple-occupancy. The Hamiltonian can be written in terms of the Hubbard operators,

$$H = \sum_{\sigma=1,2,\ldots,N} \sum_{i \neq j} t_{ij} X_i^{\sigma 0} X_j^{0\sigma}. \tag{2}$$

Let us denote the number of holes by Q, that of the fermions of the first flavor by M_1, that of the second flavor by M_2, ..., that of the Nth flavor by M_N. Following notations used in previous literature, states in the Hilbert space can be represented by spin and hole excitations from the state full of fermions with the Nth flavor $|P\rangle$, as

$$|\Phi\rangle = \sum_{(\alpha,i),j} \Phi(\{x_i^{\alpha}\},\{y_j\}) \prod_{\alpha,i} b_{i\alpha}^{\dagger} \prod_{j} h_j^{\dagger} |P\rangle. \tag{3}$$

Here $b_{i\alpha}^{\dagger} = c_{i\alpha}^{\dagger} c_{iN}$ annihilates one Nth flavored fermion at site i and creats one αth flavored fermion at site i for $\alpha = 1, 2, \ldots, (N-1)$, while $h_j^{\dagger} = c_{jN}$ creats a hole at site j. Here or in the following we always implicitly assume that we work in the space of no double occupancy and no multiple occupancy in the discussion of the strong-interaction limit. The amplitude $\Phi(\{x_i^{\alpha}\},\{y_j\})$ is symmetric when exchanging the fermions at positions x_i^α and x_j^α, and antisymmetric in the hole positions $\{y_i\}$.

Let us consider the following generalized SU(N) Gutzwiller-Jastrow wave functions corresponding to uniform motion and magnetization [9],

$$\Phi_G(\{x_i^\alpha\},\{y_m\}) = \exp\frac{2\pi i}{L}\left[\sum_\alpha J_\alpha \sum_i x_i^\alpha + J_h \sum_i y_i\right]\Phi_0,$$
$$\Phi_0 = \prod_{\alpha;i<j} d^2(x_i^\alpha - x_j^\alpha)\prod_{\alpha<\beta;i,j} d(x_i^\alpha - x_j^\beta)\prod_{\alpha,i,m} d(x_i^\alpha - y_m)\prod_{m<n} d(y_m - y_n) \tag{4}$$

where $\alpha, \beta = 1, 2, \ldots, N-1$. The quantum numbers J_α and J_h govern the momenta of the fermions and the holes. They can be integers or half integers, such that the wave functions are periodic (or antiperiodic) for odd L (for even L) under the translations $x_i^\alpha \to x_i^\alpha + L$, and $y_m \to y_m + L$. For the wave functions to be eigenstates of the Hamiltonain, the quantum numbers must be choosen from some restricted regions, which are to be specified below.

To demonstrate that the wave functions are eigenstates of the Hamiltonian, we have to consider the effect of the hopping operator very carefully. The hopping operator can be broken into N parts, each corresponding to the hopping of fermions of different flavors. Let us first consider the hopping operator of the Nth flavor, $\hat{T}(N) = \sum_{i\neq j} t_{ij} c_{iN}^\dagger c_{jN}$. When it operates on the wave functions, the hopping of the fermions of Nth flavor is equivalent to the hopping of holes

$$\frac{\hat{T}(N)\Phi_G}{\Phi_G} = -it\sum_{\bar{n}=1}^{L-1}\frac{(-1)^{\bar{n}}}{d(\bar{n})}z^{\bar{n}J_h}\sum_n\prod_{m\neq n}F_{nm}(\bar{n})\prod_{(\alpha,i)}F_{n(\alpha,i)}(\bar{n}), \tag{5}$$

where

$$F_{nm}(\bar{n}) = \cos\frac{\bar{n}\pi}{L} + \sin\frac{\bar{n}\pi}{L}\cot\Theta_{nm},$$
$$F_{n(\alpha i)}(\bar{n}) = \cos\frac{\bar{n}\pi}{L} + \sin\frac{\bar{n}\pi}{L}\cot\Theta_{n(\alpha i)};$$
$$\Theta_{nm} = \pi(y_n - y_m)/L, \qquad \Theta_{n(\alpha i)} = \pi(y_n - x_i^\alpha)/L.$$

The sum can be carried out after expanding the products and classifying terms by the number of particles involved. In the end, only the zero-particle term and two-particle terms are left. Many particle terms vanish, yielding the following result:

$$\frac{\hat{T}(N)\Phi_G}{\Phi_G} = -\frac{2\pi t}{L}QJ_h + (2\pi t/L)i\sum_n\sum_{\alpha,i}\cot\Theta_{n(\alpha i)}. \tag{6}$$

This result is valid under the condition

$$|J_h| \leq L/2 - [Q + (M_1 + M_2 + \cdots + M_{N-1})]/2.$$

To consider the effects of other parts of the hopping operators, we cannot use the wave functions directly, since the hopping will involve the fermions and holes simultaneously when they operate on the wave functions. We can generalize the idea of the spin-rotated version developed in the recent work of the $1/r^2$ t-J model to this SU(N) case. For example, to deal with the hopping operator $T(N-1) = \sum_{i\neq j} c^\dagger_{i(N-1)} c_{j(N-1)}$, we can write the Gutzwiller-Jastrow wave functions in terms of the hole positions and the positions of the fermions of flavors excluding the $(N-1)$th flavor. In terms of these coordinates, the wave functions can be found to be still in a similar product form, and thus the effect of the operator $T(N-1)$ can be calculated in the same way as for $T(N)$.

For the other hopping operators $T(N-2), T(N-3), \ldots, T(2), T(1)$, similar procedures can be carried out. After adding all the effects of the hopping operators together, the two-particle terms vanish since positions of all the fermions and the positions of the holes span the entire lattice. Thus the Gutzwiller-Jastrow wave functions are found to be exact eigenstates of the Hamiltonian, with eigenenergies given by

$$E(J_h; J_1, J_2, \ldots, J_{N-1}) = -(2\pi t/L)Q[J_h + \tilde{J}_h^{(1)} + \tilde{J}_h^{(2)} + \cdots + \tilde{J}_h^{(N-1)}], \quad (7)$$

where we have

$$\begin{aligned} \tilde{J}_h^{(1)} &= J_h - J_{N-1} + L/2, \\ \tilde{J}_h^{(2)} &= J_h - J_{N-2} + L/2, \ldots, \\ \tilde{J}_h^{(N-1)} &= J_h - J_1 + L/2. \end{aligned}$$

The many-particle terms vanish, and thus our result holds, under the conditions:

$$\begin{aligned} &|J_h| \leq (M_N)/2, \\ &|\tilde{J}_h^{(1)}| \leq (M_{N-1})/2, \\ &|\tilde{J}_h^{(2)}| \leq (M_{N-2})/2, \\ &\ldots \\ &|\tilde{J}_h^{(N-1)}| \leq (M_1)/2. \end{aligned} \quad (8)$$

Here the ground state energy is given by $E_0 = -(2\pi|t|/L)[L/2 - Q/2]Q$.

For this multiple-component system, the spectrum can also be written in terms of more generalized Jastrow functions. Here, we just write down the spectrum without getting into the detailed algebra as follows:

$$E = -(2\pi t/L)\sum_{i=1}^{Q} K_i + (\pi t Q/L)(L+1), \quad (9)$$

where K_i takes values from the region $(1, 2, \dots, L)$. Each energy level is determined by a charge configuration such as (101010) for $Q = 3$ and $L = 6$, where the 1's represent the values occupied by the charge momenta K_i. In this system, the spin and charge degrees are decoupled from each other in the entire Hilbert space. On these physical grounds, we see that for each charge configuration, the degeneracy of the corresponding energy level is given by the number of the ways to distribute the free spins among the $L - Q$ empty values. With this result, we find the free energy per lattice site given by

$$F(T, \mu)/L = -\mu - \frac{T}{2\pi} \int_{-\pi}^{\pi} dq \, \ln[N + e^{\beta(qt-\mu)}], \tag{10}$$

where μ is the chemical potential of the fermions. This free energy has also been found to correctly reproduce the first three terms in the high-temperature perturbation expansion.

In summary, we have solved the multiple-component Hubbard model in the strong-interaction limit. In this limit, the spin degrees of freedom decouple from the charge degrees of freedom in the *entire* excitation spectrum, and the system is a special example of the SU(N) Luttinger liquids in the sense of Haldane. We have shown that the SU(N) Gutzwiller-Jastrow wave functions are eigenstates of the Hamiltonian.

In the end, we notice that in the half filling and large U limit, our model reduces to the SU(N) Haldane-Shastry spin model with $1/r^2$ exchange interaction. We suspect that our multiple-component Hubbard model of the $1/r$ hopping is also completely integrable for *arbitrary* on-site energy U at *arbitrary* filling numbers. However, we have not found any elegant way to obtain the wave functions and the energy spectrum for the finite on-site energy case. It is very likely that the SU(N) system also exhibits a metal-insulator phase transition at half filling when changing the bandwidth and the on-site energy, as in the SU(2) case discovered by Gebhard and Ruckenstein about two years ago. It is also of great interest to study the ground-state properties of the system as a function of the interaction strength, such as the spin and charge susceptibilities and various ground-state correlators. It also remains to find the integrability condition for the model at the finite on-site energy.

Acknowledgments

This work was supported in part by the National Science Foundation under Grant No. NSF-DMR-89-13692. The author wishes to thank Professor P. Coleman for fruitful discussions and constant encouragment during this work, and is also grateful to him for reading the manuscript and for useful comments.

References

[1] D. Arovas and S. M. Girvin (unpublished).

[2] A. Auerbach, D. Arovas, and D. Haldane, Phys. Rev. Lett. **60**, 531 (1988).

[3] C. S. Hellberg and E. J. Mele, Phys. Rev. B **44**, 1360 (1991); Phys. Rev. Lett. **67**, 2080 (1991); *Lecture Notes in Physics*, edited by B. S. Shastry *et al.* (Spring-Verlag, Berlin, 1985), Vol. 242, p. 1.

[4] F. D. M. Haldane, Phys. Rev. Lett. **60**, 635 (1988).

[5] B. S. Shastry, Phys. Rev. Lett. **60**, 639 (1988).

[6] F. D. M. Haldane, Phys. Rev. Lett. **66**, 1529 (1991).

[7] Y. Kuramoto and H. Yokoyama, Phys. Rev. Lett. **67**, 1338 (1991).

[8] N. Kawakami, Phys. Rev. B **46**, 3191 (1992); **46**, 1005 (1992).

[9] N. Kawakami, Phys. Rev. B **45**, 7525 (1992); **47**, 2928 (1993).

[10] B. Sutherland, Phys. Rev. A **5**, 1372 (1972); **4**, 2019 (1971); J. Math. Phys. **12**, 251 (1971); **12**, 246 (1971),

[11] F. Gebhard and A. E. Ruckenstein, Phys. Rev. Lett. **68**, 244 (1992).

[12] P. W. Anderson, B. S. Shastry, and D. Hristopulos, Phys. Rev. B **40**, 8939 (1989).

[13] We note that infinite sets of conserved quantities have been found for some closely related models. See A. P. Polychronakos, Phys. Rev. Lett. **69**, 703 (1992); **70**, 2329 (1993); M. Folwer and J. A. Minahan, *ibid.* **70**, 2325 (1993).

[14] Z. Ha and F. D. M. Haldane, Phys. Rev. **46**, 9359 (1992).

[15] D. F. Wang, James T. Liu, and P. Coleman, Phys. Rev. B **46**, 6639 (1992).

[16] D. F. Wang, Q. F. Zhong, and P. Coleman (unpublished).

[17] B. Sutherland and B. Sriram Shastry, Phys. Rev. Lett. **71**, 5 (1993).

[18] B. Sriram Shastry and B. Sutherland (unpublished).

[19] Holger Frahm (unpublished).

INVARIANTS OF THE $1/r^2$ SUPERSYMMETRIC t-J MODELS

D. F. Wang
and C. Gruber
Institut de Physique Théorique
École Polytechnique Fédérale de Lausanne
PHB-Ecublens, CH-1015 Lausanne
Switzerland

Abstract In this work, we have studied the invariants of motion of two SU(N) supersymmetric t-J models of $1/r^2$ hopping and exchange in one dimension. The first model is defined on a lattice of equal spaced sites, and the second on a nonequal spacing lattice. Using the "exchange operator formalism", we are able to construct all the invariants for the models, by mapping the systems to mixtures of fermions and bosons. This identification shows that the supersymmetric t-J model on the chain with equal-spaced sites also belongs to Shastry-Sutherland's "super-lax-pair" family.

Since the independent works by Haldane and Shastry, there have been renewed interests in exactly solvable models of long-range interaction [1, 2, 3, 4, 5, 6]. Of these systems, the one-dimensional (1D) supersymmetric t-J model of $1/r^2$ exchange and hopping has been studied intensively [6, 7, 8, 14]. The system is identified as a free Luttinger liquid [6, 7, 8], and the asymptotic correlation functions have been calculated through the finite-size-scaling technique [7]. The excitation spectrum of the system may be obtained with the help of the asymptotic Bethe ansatz [7]. In particular, for the SU(2) case, the asymptotic Bethe ansatz spectrum was explicitly shown to be exact, and the correct thermodynamics was given when the spinon rotation was properly taken into account [8]. In general, exact solvability implies the existence of infinite number of constants of motion. For the long-range t-J models, the complete construction of invariants of motion has remained unknown. In this work, applying the "exchange operator formalism" to a mixture of fermions and bosons, we are able to systematically provide all the invariants for the SU(N) systems.

Let us first consider the $1/r^2$ supersymmetric t-J model on a one dimensional lattice of equal spaced sites. The Hamiltonian for the one-dimensional t-J

Reprinted from Wang *et al.*, Phys. Rev. B 49 (1994) 15712-15714

model is given by

$$H = P_G\Bigg[-\sum_{1\le i\neq j\le L}\sum_{\sigma=1}^{N} t_{ij}(c_{i\sigma}^\dagger c_{j\sigma}) + \sum_{1\le i\neq j\le L} J_{ij}\left[P_{ij}-(1-n_i)(1-n_j)\right]\Bigg]P_G, \qquad (1)$$

where we take the hopping matrix and the spin exchange interaction to be $t_{ij}/2 = J_{ij} = 1/d^2(i-j)$, and $d(n) = (L/\pi)\sin(n\pi/L)$ is the chord distance, with L the size of the lattice. The operator $c_{i\sigma}^\dagger$ is the fermionic operator to create an electron with spin component σ at site i, $c_{i\sigma}$ is the corresponding fermionic annihilation operator. Their anticommutation relations are given by $\{c_{i\sigma_i}, c_{j\sigma_j}^\dagger\}_+ = \delta_{ij}\delta_{\sigma_i\sigma_j}$, $\{c_{i\sigma_i}, c_{j\sigma_j}\}_+ = 0$, $\{c_{i\sigma_i}^\dagger, c_{j\sigma_j}^\dagger\}_+ = 0$. We assume that the spin component σ takes values from 1 to N. The Gutzwiller projection operator P_G projects out all the double or multiple occupancies, $P_G = \prod_{i=1}^{L} P_G(i)$, and $P_G(i) = \delta_{0,n_i} + \delta_{1,n_i}$, with $n_i = \sum_{\sigma=1}^{N} c_{i\sigma}^\dagger c_{i\sigma}$. The operator

$$P_{ij} = \sum_{\sigma=1}^{N}\sum_{\sigma'=1}^{N} c_{i\sigma}^\dagger c_{i\sigma'} c_{j\sigma'}^\dagger c_{j\sigma}$$

exchanges the spins of the electrons at sites i and j, if both sites are occupied. n_i and n_j are the electron number operators at sites i and j.

Now, on the lattice, we may introduce two new fields, the f and b fields. For the new fields, we have $\{f_{i\sigma}, f_{j\sigma'}\}_+ = 0$, $\{f_{i\sigma}, f_{j\sigma'}^\dagger\}_+ = \delta_{ij}\delta_{\sigma\sigma'}$, $[b_i, b_j] = 0$, $[b_i, b_j^\dagger] = \delta_{ij}$. The b fields always commute with the f fields. The size of the Hilbert space at each site is ∞ in this case. However, let us project out the zero occupancy and all the double or multiple occupancies, and work in the subspace where there is exactly one particle at each site. This new subspace can be shown to be equivalent to the subspace defined by the c field with no double or multiple occupancies. In particular, we may represent the fermionic electron operators $c_{i\sigma}^\dagger$ and $c_{i\sigma}$ in the following way:

$$\begin{aligned} P_G(i)c_{i\sigma}^\dagger P_G(i) &= \delta_{1,n_b^i+n_f^i} f_{i\sigma}^\dagger b_i \delta_{1,n_b^i+n_f^i}, \\ P_G(i)c_{i\sigma} P_G(i) &= \delta_{1,n_b^i+n_f^i} b_i^\dagger f_{i\sigma} \delta_{1,n_b^i+n_f^i}, \end{aligned} \qquad (2)$$

where $n_b^i + n_f^i = b_i^\dagger b_i + \sum_{\sigma=1}^{N} f_{i\sigma}^\dagger f_{i\sigma}$. In terms of the f and b fields, a state vector can be written as

$$\begin{aligned} |\phi\rangle &= \sum_{\sigma_1,\sigma_2,\ldots,\sigma_{N_e}} \sum_{\{x\},\{y\}} \phi(x_1\sigma_1, x_2\sigma_2, \ldots, x_{N_e}\sigma_{N_e} | y_1, y_2, \ldots, y_Q) \\ &\quad \times f_{x_1\sigma_1}^\dagger f_{x_2\sigma_2}^\dagger \cdots f_{x_{N_e}\sigma_{N_e}}^\dagger b_{y_1}^\dagger b_{y_2}^\dagger \cdots b_{y_Q}^\dagger |0\rangle, \end{aligned} \qquad (3)$$

where N_e is the number of f fermions on the lattice, Q is the number of b bosons, and we require that $x_i \neq x_j \neq y_k \neq y_l$, and that the f fermion positions $\{x\}$ and the b boson positions $\{y\}$ span the whole chain. Obviously, N_e is also the number of electrons, and Q is also the number of holes on the lattice. The amplitude ϕ is antisymmetric when exchanging $(x_i\sigma_i)$ and $(x_j\sigma_j)$, and symmetric in the boson coordinates $\{y\} = (y_1, y_2, \cdots, y_Q)$. Using the mapping Eq. (2) in a straightforward way, the Hamiltonian of the supersymmetric t-J model can be written in terms of the fermionic f field and the bosonic b field.

With the above mapping, we may write the eigenenergy equation of the supersymmetric t-J model in the first quantized form. Define the "exchange operator" M_{ij} as

$$M_{ij}F(q_1, q_2, \ldots, q_i, \ldots, q_j, \ldots, q_L) = F(q_1, q_2, \ldots, q_j, \ldots, q_i, \ldots, q_L),$$

where the function F is an arbitrary function of some position variables $(q_1, q_2, \cdots, q_L)$, i.e., the operator M_{ij} exchanges the positions q_i, q_j of the particles i and j. In terms of such exchange operators, the eigenenergy equation of the t-J model takes the form as follows [17]:

$$-\left[\sum_{1\leq i\neq j\leq L} d^{-2}(q_i - q_j)M_{ij}\right]\phi(\{q\};\{\sigma\}) = E\phi(\{q\};\{\sigma\}), \qquad (4)$$

where

$$\{q\} = (q_1, q_2, \ldots, q_L) = (x_1, x_2, \ldots, x_{N_e}, y_1, y_2, \ldots, y_Q)$$

and

$$\begin{aligned}\phi(\{q\};\{\sigma\}) &= \phi(q_1\sigma_1, q_2\sigma_2, \ldots, q_{N_e}\sigma_{N_e}|q_{N_e+1}q_{N_e+2}\cdots q_L) \\ &= \phi(x_1\sigma_1, x_2\sigma_2, \ldots, x_{N_e}\sigma_{N_e}|y_1, y_2, \ldots, y_Q)\end{aligned}$$

is the amplitude of the state vector of Eq. (3). $\{\sigma\} = (\sigma_1, \sigma_2, \ldots, \sigma_{N_e})$ are the spin variables of the f fermions. The operation M_{ij} is defined in the conventional way:

$$M_{ij}\phi(\{q\};\{\sigma\}) = \phi(\{q'\};\{\sigma\}),$$

with

$$\{q\} = (q_1, q_2, \ldots, q_i, \ldots, q_j, \ldots, q_L)$$

and

$$\{q'\} = (q_1, q_2, \ldots, q_j, \ldots, q_i, \ldots, q_L).$$

Here, the sum in the Eq. (4) is over all pairs of particles. Thus, we see that using the f and b fields, we can write the original t-J model as an eigenvalue problem for a mixture of the f fermions and the spinless b bosons in terms of the "exchange operators".

Recently, Fowler and Minahan have considered a gas of identical bosons on a one-dimensional chain [9]. Using the so-called "exchange operator formalism" [10], they have been able to construct explicitly all the invariants of motion for the SU(N) spin chain of Haldane and Shastry. Let us briefly review their results. Say M_{ij} is the exchange operator that interchanges q_i and q_j, the positions of the particle i and particle j, when operating M_{ij} on a wave function $F(q_1, q_2, \ldots, q_L)$. In terms of this operator, they have been able to construct an infinite set of quantities I_n that commute among themselves:

$$[I_n, I_m] = 0, \tag{5}$$

where $I_n = \sum_{i=1}^{L} \pi_i^n$, with $\pi_i = \sum_{j(\neq i)} (z_j/z_{ij}) M_{ij}$, $z_i = e^{2\pi i q_i/L}$, $z_{ij} = z_i - z_j$, and $n, m = 0, 1, 2, \ldots, \infty$. It was found that all these quantities commute with the Hamiltonian $H = \sum_{1 \leq i \neq j \leq L} |z_i - z_j|^{-2} M_{ij}$ as long as the particles occupy the whole chain. For a system of identical bosons on the chain, the wave function is totally symmetric when we simultaneously interchange spins and positions of two particles. The effect of the exchange operator M_{ij} is just equivalent to the effect of the spin exchange operator alone. Using this method, they have successfully constructed all the invariants for the SU(N) Haldane-Shastry model.

We would like to stress that, in the language of the exchange operators M_{ij}, the commutation results proved by Fowler and Minahan hold as operator identities. The central issue is that the form of the wave functions of many-particle systems, as well as the statistics of the particles or the types of particles, do *not* matter in order for the commutators to hold, as long as the particles occupy the whole chain. We may then apply the "exchange operator formalism" to the wave functions of mixtures of fermions and bosons. Therefore, from the eigenequation, Eq. (4), we conclude that in the first quantization all the invariants of the t-J model are the same I_n's as constructed by Fowler and Minahan, which can be written in terms of the exchange operators M_{ij}'s.

With the permutation properties of the amplitude ϕ for the mixture of bosons and fermions, it is straightforward to write all the invariants of motion of the t-J model in the second quantization form using the I_n's. For instance, the exchange operation between the f fermion positions is equivalent to the spin exchange operation (minus sign involved), the exchange operation between b boson positions is equivalent to the hole-hole interaction term, and exchange operation between f fermion and b boson positions is equivalent to the electron hopping. Such a procedure to reduce an I_n to a second quantized form is quite simple, and we do not write all the details. Thus we provide a systematic way to construct all the invariants of motion for the $1/r^2$ supersymmetric t-J model, either in first quantized or in second quantized forms.

Recently, Shastry and Sutherland have studied the interesting relation between supersymmetry and integrability, through the so-called "super-lax-pair" approach [4, 5]. For this equal-spacing chain, using the mapping Eq. (2), we have been able to write the t-J model Hamiltonian, Eq. (1), in terms of the exchange operators as Eq. (4). This identification shows that the "super-lax-pair" results obtained by Shastry and Sutherland may apply to this t-J model [4, 5].

Besides the above integrable t-J model on equally spaced sites, let us consider another supersymmetric t-J model of $1/r^2$ hopping and exchange on a chain with sites not equally spaced. The positions of the sites $x_1, x_2, \ldots, x_L$ are determined by the equation

$$x_i = \sum_{1 \le j (\neq i) \le L} 2/(x_i - x_j)^3. \tag{6}$$

This equation has appeared before in a paper discussing a long-range spin chain of Haldane-Shastry type [10]. Doping this spin chain, we are led to the following supersymmetric t-J model:

$$\begin{aligned} H &= P_G \Bigg[- \sum_{1 \le i \neq j \le L} \sum_{\sigma=1}^{N} t_{ij} (c_{i\sigma}^{\dagger} c_{j\sigma}) \\ &\quad + \sum_{1 \le i \neq j \le L} J_{ij} \left[P_{ij} - (1 - n_i)(1 - n_j) \right] \Bigg] P_G, \end{aligned} \tag{7}$$

where the hopping matrix and the antiferromagnetic exchange interaction are given by $t_{ij}/2 = J_{ij} = 1/(x_i - x_j)^2$, and each site is occupied at most by one electron.

In the half-filled case $N_e = L$, this system reduces to the spin chain that has been studied before, which is completely solvable and a similar exchange operator formalism has been developed [10, 11]. Let us just write down the results obtained by Polychronakos: $[I_n, I_m] = 0$, $[I_n, H] = 0$, where $I_n = \sum_{i=1}^{L} h_i^n$, $h_i = a_i^{\dagger} a_i$, and $a_i^{\dagger} = \pi_i^{\dagger} + i q_i$, $a_i = \pi_i - i q_i$, with $\pi_i = \sum_{j(\neq i)} i(q_i - q_j)^{-1} M_{ij}$, $H = \sum_{i \neq j} (q_i - q_j)^{-2} M_{ij}$, and $n, m = 0, 1, 2, \ldots, \infty$. Here, all the particles are put on the chain where the sites are positioned as determined by Eq. (6). We may relate the operation of exchanging particle positions to the operation of exchange particle spins, by assuming that we have identical bosons again, for which the wave functions are totally symmetric when we exchange the spins and positions of two particles simultaneously [10]. With this assumption, from I_n's, we thus can derive all the invariants of motion written in terms of the spin exchange operators alone.

Again, all commutation results written in terms of the exchange operators M_{ij} obtained by Polychronakos hold as operator identities, as long as the particles occupy the whole chain of the sites positioned in the special way. The forms of the wave functions do *not* matter at all. Thus, the commutation results can be applied to wave functions of particles of arbitrary statistics or wave functions of mixtures of particles of different statistics on the chain. Mapping our supersymmetric t-J model in terms of the b and f fields, we can also write the eigenenergy equation in first quantized form. In terms of the exchange operators between the positions of the bosons and fermions, the Hamiltonian takes the form

$$H = - \sum_{1 \leq i \neq j \leq L} (q_i - q_j)^{-2} M_{ij}. \tag{8}$$

Applying the formalism to this t-J model, in a similar way we obtain all the invariants, either in first quantized or in the second quantized form, which commute among themselves and with the Hamiltonian. Thus this supersymmetric t-J model is also completely integrable.

In conclusion, we have studied the invariants of motion of two SU(N) supersymmetric t-J models of long-range hopping and exchange. The first system is on the chain of equal-spaced sites, and the other on a chain of nonequal-spaced sites. Mapping the corresponding t-J model Hamiltonians to those written in terms of mixed fermionic and bosonic fields, then applying the "exchange operator formalism", we were able to construct systematically all the invariants of the original Hamiltonian.

Finally, we wish to point out that, the second t-J model has also a metal-insulator phase transition at half filling. Away from half filling, we expect to have decoupled spin and charge excitations near the ground state. The system would be a Luttinger liquid. The study of the physical properties of the t-J model, such as its full excitation spectrum, is reported in our forthcoming paper. We obtain Jastrow product ground-state and excited-state wave functions, as in the case for the model on the chain with equally spaced sites. It would also be very interesting to find out possible Shastry-Sutherland type "super-lax-pair" for this supersymmetric t-J model on a nonequal-spacing chain. We will return to the issue of constructing invariants of the nonsupersymmetric t-J models with $1/r^2$ hopping and exchange in future.

Acknowledgments

This work was supported by the World Laboratory.

References

[1] F. D. M. Haldane, Phys. Rev. Lett. **60**, 635 (1988).

[2] B. S. Shastry, Phys. Rev. Lett. **60**, 639 (1988); **69**, 164 (1992).

[3] F. D. M. Haldane, Phys. Rev. Lett. **66**, 1529 (1991).

[4] B. S. Shastry and B. Sutherland, Phys. Rev. Lett. **70**, 4092 (1993).

[5] B. Sutherland and B. S. Shastry, Phys. Rev. Lett. **72**, 5 (1993).

[6] Y. Kuramoto and H. Yokoyama, Phys. Rev. Lett. **67**, 1338 (1991).

[7] N. Kawakami, Phys. Rev. B **45**, 7525 (1992); **46**, 1005 (1992).

[8] D. F. Wang, James T. Liu, and P. Coleman, Phys. Rev. B **46**, 6639 (1992).

[9] M. Fowler and J. A. Minahan, Phys. Rev. Lett. **70**, 2325 (1993).

[10] A. P. Polychronakos, Phys. Rev. Lett. **70**, 2329 (1993); **69**, 703 (1992).

[11] Holger Frahm, J. Phys. A **26**, 473 (1993).

[12] B. Sutherland, Phys. Rev. A **5**, 1372 (1972); **4**, 2019 (1971); J. Math. Phys. **12**, 251 (1971); **12**, 246 (1971).

[13] F. Gebhard and A. E. Ruckenstein, Phys. Rev. Lett. **68**, 244 (1992).

[14] Z. N. C. Ha and F. D. M. Haldane, Phys. Rev. B **46**, 9359 (1992).

[15] D. F. Wang, Q. F. Zhong and P . Coleman, Phys. Rev. B **48**, 8476 (1993).

[16] D. F. Wang, Phys. Rev. B **48**, 10 556 (1993).

[17] As in Ref. [5], Eq. (4) identifies the t-J model with the intrinsic dynamic pairs, in the strong interaction limit, of the integrable continuous system $H = \frac{1}{2}\sum_{i=1}^{L} p_i^2 + \sum_{i<j}[1(1 - M_{ij})]/[d^2(x_i - x_j)]$ for a mixture of Q spinless bosons and N_e fermions with spins.

EXACT RESULTS OF THE ONE-DIMENSIONAL $1/r^2$ SUPERSYMMETRIC t-J MODEL WITHOUT TRANSLATIONAL INVARIANCE

C. Gruber
and D. F. Wang
Institut de Physique Théorique
École Polytechnique Fédérale de Lausanne
PHB-Ecublens, CH-1015 Lausanne
Switzerland

Abstract In this work, we continue the study of the supersymmetric t-J model with $1/r^2$ hopping and exchange without translational invariance. A set of Jastrow eigenfunctions are obtained for the system, with eigenenergies explicitly calculated. The ground state of the t-J model is included in this set of wave functions. The spectrum of this t-J model consists of equidistant energy levels which are highly degenerate.

In recent years, there have been considerable interests in study of low-dimensional electronic models of strong correlation, due to the possibility that the normal state of the two-dimensional (2D) novel superconductivity [1] may share some interesting feature of a 1D interacting electron system (non-Fermi-liquid behavior). The one-band two-dimensional Hubbard model reduces to the t-J model in the large on-site energy limit. The Hubbard model and the t-J model have been under intense study through various approaches. For these strongly correlated electron models, few exact results may be obtained in two dimensions. The high-dimensional versions are much harder to study than their one-dimensional ones. In one dimension, however, the Bethe-ansatz technique may allow us to exactly solve Hamiltonians in some special cases, such as the Lieb-Wu solution [2] and the ordinary t-J model at its supersymmetric point [3]. In particular, the 1D long-range exactly solvable electronic models have attracted a lot of attention, since they display interesting physics with solutions of simple mathematical structure [4, 5, 6, 7, 8, 9, 10, 11, 12, 13, 14, 15, 16, 17, 18, 19, 20, 21].

Reprinted from Gruber *et al.*, Phys. Rev. B 50 (1994) 3103-3107

Recently, we have been able to explicitly construct all the constants of motion for the translational invariant long-range supersymmetric t-J model, by mapping the system to a mixture of fermions and bosons, with the superalgebra representation for the electron fields [12]. Moreover, we have introduced a one-dimensional supersymmetric t-J model with $1/r^2$ hopping and exchange without translational invariance. This system has also been shown by us to be completely integrable, with infinite number of conserved quantities explicitly constructed [12]. In this work, we continue the study of this t-J model. A set of Jastrow eigenfunctions, as well as their eigenenergies, are obtained explicitly. The ground state of the t-J model is included in this set of wave functions. We also briefly discuss the structure of the full spectrum for the system.

The Hamiltonian for the one-dimensional t-J model is given by [12]

$$\begin{aligned} H_{t\text{-}J} &= (1/2)P_G\Bigg[-\sum_{1\le i\ne j\le L}\sum_{\sigma=1}^{N} t_{ij}(c_{i\sigma}^{\dagger}c_{j\sigma}) \\ &\qquad + \sum_{1\le i\ne j\le L} J_{ij}\left[P_{ij}-(1-n_i)(1-n_j)\right]\Bigg]P_G, \end{aligned} \tag{1}$$

where we take the hopping matrix and the spin exchange interaction to be $t_{ij}/2 = J_{ij} = 1/(r_i - r_j)^2$. L is the number of sites on the chain. In this model, the positions of the sites $\{r_i\}$ are given by the roots of the Hermite polynomial $H_L(x)$, and the spin component σ takes values from 1 to N. The operator $c_{i\sigma}^{\dagger}$ is the the creation operator for an electron at site i with spin σ; $c_{i\sigma}$ is the corresponding annihilation operator. Their anticommutation relations are given by $\{c_{i\sigma_i}, c_{j\sigma_j}^{\dagger}\}_+ = \delta_{ij}\delta_{\sigma_i\sigma_j}$, $\{c_{i\sigma_i}, c_{j\sigma_j}\}_+ = 0$, $\{c_{i\sigma_i}^{\dagger}, c_{j\sigma_j}^{\dagger}\}_+ = 0$. The Gutzwiller projection operator P_G projects out all the double or multiple occupancies, $P_G = \prod_{i=1}^{L} P_G(i)$, and $P_G(i) = \delta_{0,n_i} + \delta_{1,n_i}$, with $n_i = \sum_{\sigma=1}^{N} c_{i\sigma}^{\dagger}c_{i\sigma}$. The operator $P_{ij} = \sum_{\sigma=1}^{N}\sum_{\sigma'=1}^{N} c_{i\sigma}^{\dagger}c_{i\sigma'}c_{j\sigma'}^{\dagger}c_{j\sigma}$ exchanges the spins of the electrons at sites i and j, if both sites are occupied, and is zero otherwise. At half-filling, our t-J model becomes the long-range spin model, introduced first by Polychronakos on such a nontranslational-invariant lattice [14].

In terms of the b and f fields, the eigenequation of the t-J model can be written as [12]

$$-\sum_{1\le i<j\le L}(q_i-q_j)^{-2}M_{ij}\phi(\{q\},\{\sigma\}) = E\phi(\{q\},\{\sigma\}), \tag{2}$$

where $\phi(\{q\},\{\sigma\}) = \phi(q_1\sigma_1, q_2\sigma_2, \ldots, q_{N_e}\sigma_{N_e}|q_{N_e+1}, q_{N_e+2}, \ldots, q_L)$ is the amplitude for the f fermions to be at $q_1, q_2, \ldots, q_{N_e}$, while the spinless b bosons are at $q_{N_e+1}, q_{N_e+2}, \ldots, q_L$. Here, $\{\sigma\} = (\sigma_1, \sigma_2, \ldots, \sigma_{N_e})$ and $\{q\} = (q_1, q_2, \ldots, q_L) = (x_1, x_2, \ldots, x_{N_e}, y_1, y_2, \ldots, y_Q)$. The wave function ϕ is symmetric in the b boson positions $\{y\}$, while antisymmetric when exchanging

$x_i\sigma_i$ and $x_j\sigma_j$. The operator M_{ij} exchanges only the position variables q_i and q_j: $M_{ij}\phi(\{q\},\{\sigma\}) = \phi(\{q'\},\{\sigma\})$, with $\{q\} = (q_1, q_2, \ldots, q_i, \ldots, q_j, \ldots, q_L)$ and $\{q'\} = (q_1, q_2, \ldots, q_j, \ldots, q_i, \ldots, q_L)$. In this approach, the f fermions and the b bosons occupy the whole chain; i.e., $\{q\}$ and $\{q'\}$ are permutations of the sites $\{r_1, r_2, \ldots, r_L\}$, and we work in the Hilbert space where at each site there is exactly one particle. Q, the number of the b bosons, is also the number of holes in the original problem; N_e, the number of the f fermions, is also the number of the c electrons on the lattice. $\tilde{N}_\alpha$, with $\alpha = 1, 2, \ldots N$, the number of the f fermions with spin component α, is also the number of the c electrons with spin component α on the chain.

The Hamiltonian in the first quantization, as given by Eq. (2), is

$$H = -\sum_{1\le i<j\le L} (q_i - q_j)^{-2} M_{ij}. \tag{3}$$

It commutes with the permutation operator $T_{ij} = P_{ij}^{\sigma} M_{ij}$ exchanging the f fermion spin and position simultaneously. Let us work in the Hilbert space where the number of fermions of each flavor is fixed, i.e., $\tilde{N}_\sigma$, $\sigma = 1, 2, \ldots, N$, is fixed. Consider the following wave function in Jastrow product form:

$$\phi(x_1\sigma_1, x_2\sigma_2, \ldots, x_{N_e}\sigma_{N_e} | y_1, y_2, \ldots, y_Q) = \prod_{i<j} (x_i - x_j)^{\delta_{\sigma_i\sigma_j}} e^{i\frac{\pi}{2}\mathrm{sgn}(\sigma_i - \sigma_j)}, \tag{4}$$

where $\{x\}$ and $\{y\}$ span the whole lattice. We would like to show that this wave function is an eigenstate of the system. The Hamiltonian in Eq. (2) can be broken up into three parts: The first part H_1 exchanges the f fermions, the second H_2 exchanges the b bosons, and the third H_3 exchanges the bosons and the fermions:

$$\begin{aligned} H_1 &= (-1) \sum_{1\le i<j\le N_e} (q_i - q_j)^{-2} M_{ij}, \\ H_2 &= (-1) \sum_{1+N_e\le\alpha<\beta\le L} (q_\alpha - q_\beta)^{-2} M_{\alpha\beta}, \\ H_3 &= (-1) \sum_{N_e+1\le\alpha\le L}\ \sum_{1\le j\le N_e} (q_\alpha - q_j)^{-2} M_{\alpha j}. \end{aligned} \tag{5}$$

We then calculate the effects of these three parts when acting the Jastrow wave function given by Eq. (4). The contribution from H_2 is immediate:

$$H_2\phi = -\sum_{\alpha<\beta} (y_\alpha - y_\beta)^{-2}\phi. \tag{6}$$

The contributions from H_1 and H_3 are harder to deal with since many particle terms are involved. Using a similar trick introduced in Refs. [8, 19], we have

$$H_3\phi \;=\; -\sum_i \sum_\alpha (x_i - y_\alpha)^{-2} \prod_{j(\neq i)} \left(\frac{y_\alpha - x_j}{x_i - x_j}\right)^{\delta_{\sigma_i\sigma_j}} \phi$$

$$= -\sum_i \sum_\alpha (x_i - y_\alpha)^{-2} \prod_{j(\neq i)} (1 + \delta_{\sigma_i\sigma_j} \frac{y_\alpha - x_i}{x_i - x_j})\phi$$

$$= -\sum_i \sum_\alpha \left[(x_i - y_\alpha)^{-2} + \sum_{j(\neq i)} \frac{\delta_{\sigma_i\sigma_j}}{(y_\alpha - x_i)(x_i - x_j)} \right] \phi - \text{rest}, \tag{7}$$

where

$$\text{rest} = \sum_\alpha \sum_{\sigma=1}^{N} \sum_{r=3}^{N_e} \sum_{\substack{J\subset \wp^\sigma \\ |J|=r}} \left[\sum_{i\in J} (y_\alpha - x_i)^{r-3} \prod_{x_j\in X_J/x_i} \frac{1}{(x_i - x_j)} \phi \right], \tag{8}$$

with $\wp^\sigma = \{k \in \{1, 2, \ldots, N_e\}; \sigma_k = \sigma\}$, $X_J = \{x_j; j \in J\}$. Then using the fact that for any set $X = (x_1, x_2, \ldots, x_n)$, we have the identity (see Appendix A)

$$\sum_{i=1}^{n} x_i^t \prod_{x_j\in X/x_i} \frac{1}{(x_i - x_j)} = 0, \tag{9}$$

for all $t = 0, 1, 2, \ldots, n-2$, we conclude that rest $= 0$.

The contribution from H_1 is calculated in a similar manner:

$$H_1\phi = -\sum_{i<j} (x_i - x_j)^{-2} (1 - 2\delta_{\sigma_i\sigma_j}) \prod_{k(\neq i,j)} (\frac{x_k - x_j}{x_k - x_i})^{\delta_{\sigma_i\sigma_k}} (\frac{x_k - x_i}{x_k - x_j})^{\delta_{\sigma_j\sigma_k}} \phi$$

$$= \sum_{i<j} \delta_{\sigma_i\sigma_j} (x_i - x_j)^{-2}\phi - \sum_{i<j} (x_i - x_j)^{-2} (1 - \delta_{\sigma_i\sigma_j})$$

$$\times \prod_{k(\neq i,j)} \left(1 + \delta_{\sigma_i\sigma_k} \frac{x_i - x_j}{x_k - x_i} - \delta_{\sigma_j\sigma_k} \frac{x_i - x_j}{x_k - x_j} \right)$$

$$= \sum_{i\neq j} \delta_{\sigma_i\sigma_j} (x_i - x_j)^{-2}\phi - \sum_{i<j} (x_i - x_j)^{-2}\phi - \sum_{i<j} (x_i - x_j)^{-2} \times$$

$$(1 \quad - \quad \delta_{\sigma_i\sigma_j}) \sum_{\substack{K_1\subset\{1,\ldots,N_e\}/ij \\ K_2\subset\{1,\ldots,N_e\}/ij \\ K_1\cup K_2\neq 0}} \prod_{k\in K_1} \delta_{\sigma_i\sigma_k} \frac{x_i - x_j}{x_k - x_i} \cdot \prod_{k\in K_2} \frac{x_i - x_j}{x_k - x_j} (-\delta_{\sigma_j\sigma_k})\phi \tag{10}$$

Using the sum rule Eq. (9), and the fact that $\sum_{i\neq j\neq k} (x_k - x_i)^{-1} (x_k - x_j)^{-1} \delta_{\sigma_i\sigma_j} \delta_{\sigma_i\sigma_k} = 0$, the last term in the above equation becomes

$$-\phi \sum_{i<j} (1 - \delta_{\sigma_i\sigma_j}) \sum_{k(\neq i,j)} [(x_i - x_j)^{-1} (x_k - x_i)^{-1} \delta_{\sigma_i\sigma_k}$$

$$-(x_i - x_j)^{-1}(x_k - x_j)^{-1}\delta_{\sigma_j\sigma_k}]$$
$$= -\phi \sum_{i\neq j} \sum_{k(\neq i,j)} (x_i - x_j)^{-1}(x_k - x_i)^{-1}\delta_{\sigma_i\sigma_k}. \quad (11)$$

In the end, we have

$$\begin{aligned} H\phi &= -\left[\sum_{1\leq i<j\leq L} (q_i - q_j)^{-2}\right]\phi \\ &\quad + \sum_{i\neq j}(x_i - x_j)^{-1}\delta_{\sigma_i\sigma_j}\left[\sum_{k(\neq i)} (x_i - x_k)^{-1} + \sum_{\alpha}(x_i - y_\alpha)^{-1}\right]\phi \end{aligned} \quad (12)$$

Using the properties of the roots of the Hermite polynomial

$$r_i = \sum_{j(\neq i)} (r_i - r_j)^{-1}, \qquad \sum_{1\leq i<j\leq L} (r_i - r_j)^{-2} = L(L-1)/4, \quad (13)$$

we thus conclude that the wave function ϕ is an eigenstate with eigenvalue

$$E = -L(L-1)/4 + (1/2)\sum_{\sigma=1}^{N}(\tilde{N}_\sigma - 1)\tilde{N}_\sigma. \quad (14)$$

Although it is expected that this wave function is the lowest-energy state in the subspace of fixed $\tilde{N}_1, \tilde{N}_2, \ldots, \tilde{N}_N$, we were not able to prove it. However, in the case of SU(2), the small lattice diagonalization up to eight sites confirms this conjecture. Moreover, the discussion below will also confirm this idea for the general case. For fixed number N_e of the electron number, the minimum of the energy is obtained when $|\tilde{N}_\sigma - \tilde{N}_{\sigma'}|$ is as small as possible for each pair $\sigma \neq \sigma'$.

In the SU(2) case, the above result becomes

$$E = (-1)L(L-1)/4 + (1/2)\tilde{N}_\uparrow(\tilde{N}_\uparrow - 1) + (1/2)\tilde{N}_\downarrow(\tilde{N}_\downarrow - 1), \quad (15)$$

where $\tilde{N}_\uparrow$ and $\tilde{N}_\downarrow$ are the numbers of the up-spin electrons and the down-spin electrons respectively. For fixed number of electrons on the chain, i.e., for fixed N_e, the minimum of the energy given in Eq. (15) is obtained when $S_z = 0$ for even N_e, or when $S_z = \pm 1/2$ for odd N_e. Therefore, the ground state is a spin singlet (respectively spin 1/2) state for even (respectively odd) number of electrons on the chain. In particular, for an even number of electrons on the chain, the ground state energy is

$$E_G = (-1/4)L(L-1) + N_e^2/4 - N_e/2, \quad (16)$$

while for an odd number of electrons it is

$$E_G = (-1/4)L(L-1) + \left(\frac{N_e - 1}{2}\right)^2. \tag{17}$$

The charge susceptibility of the ground state χ_c is given by $\chi_c^{-1} = \partial^2 E_G/\partial N_e^2 = 1/2$, independent of the electron concentration. Very unexpectedly, the charge susceptibility is also finite at half-filling $N_e = L$, in spite of the existence of a metal-insulator phase transition at half-filling for this system. This is in contrast to the case of the periodic $1/r^2$ supersymmetric t-J model, where the charge susceptibility is divergent at half-filling, at which the metal-insulator phase transition occurs.

To study the spectrum of the system away from half-filling, we follow the idea introduced in Ref. [14]. Let us define the operators

$$\begin{aligned} \pi_j &= i \sum_{k(\neq j)} (q_j - q_k)^{-1} M_{jk} = \pi_j^\dagger, \\ a_j^\dagger &= \pi_j + i q_j, \\ a_j &= \pi_j - i q_j, \end{aligned} \tag{18}$$

which satisfy the following commutation relations:

$$\begin{aligned} [\pi_j, \pi_k] &= 0, \\ [q_j, H] &= i\pi_j, \\ [\pi_j, H] &= -2i \sum_{k(\neq j)} (q_j - q_k)^{-3}. \end{aligned} \tag{19}$$

Then using the property of the roots of the Hermite polynomial we have

$$[\pi_j, H] = -i q_j \tag{20}$$

and

$$\begin{aligned} [a_j^\dagger, H] &= -a_j^\dagger, \\ [a_j, H] &= a_j. \end{aligned} \tag{21}$$

Therefore the operators $A_i^\dagger(\nu) = a_i^\dagger S_i^{(\nu)}$, $i = 1, 2, \ldots, N_e$, where $\nu = 0, \pm, z$ for the SU(2) case with $S_i^{(0)} = 1$, will act as raising operators, while their Hermitian conjugate $A_i(\nu)$ will act as lowering operators. It thus follows that the wave function

$$\phi_{\{n\},\{\nu\}} = \sum_P \prod_{i=1}^{N_e} [A_i^\dagger(\nu_i)]^{n_{P_i}} \phi, \tag{22}$$

with $\{n\} = (n_1, n_2, \ldots, n_{N_e})$, $n_i \geq 0$, $\{\nu\} = (\nu_1, \nu_2, \ldots, \nu_{N_e})$, is either an eigenstate with energy

$$E_{\{n\}} = E + \sum_{i=1}^{N_e} n_i \tag{23}$$

or zero.

Moreover, it is shown in Appendix B that the operators $\sum_{i=1}^{N_e} A_i(\nu_i)$ with $\nu_i = 0$ or z, and a_α annihilate the wave function ϕ, and also $\sum_{i=1}^{N_e} A_i(\pm)\phi_G = 0$, confirming the conjecture that ϕ is the lowest-energy state in the subspace. We then arrive at the conjecture that the excitation spectrum of the system is of the form

$$\begin{aligned} E(s) &= E + s, \\ & s \in (0, 1, 2, \ldots, s_{\max}); \end{aligned} \tag{24}$$

i.e., the spectrum of this t-J model consists of equal-spaced energy levels. Since the model is on a finite chain, $s_{\max}$ is finite. In the special case of SU(2), the small lattice diagonalization up to eight sites suggests that the highest-energy level is given by

$$E_{\max}(Q) = L(L-1)/4 - Q(Q-1)/2; \tag{25}$$

i.e., for an even number of electrons,

$$E_{\max} = E_G + (1/4)N_e(4L - 3N_e), \tag{26}$$

where Q is the number of holes on the chain and $N_e = L - Q$ is the number of electrons. For $N_e = 1$ or $N_e = L$ (half-filling), this formula gives right results; moreover, at half-filling, this corresponds to all spins polarized in one direction.

The feature of the t-J model spectrum consisting of equal-distant energy levels may also be seen by taking the strong interaction limit of the Sutherland-Calogero-Morse quantum system for a mixture of fermions and bosons;

$$H = (-1/2)\sum_{i=1}^{L} \partial^2/\partial q_i^2 + \sum_{i=1}^{L} l^2 q_i^2/2 + \sum_{i<j} l(l - M_{ij})/(q_i - q_j)^2, \tag{27}$$

where there are N_e fermions with spins and Q spinless bosons; M_{ij} permutes the positions of the particles i and j only. The mixture gas has equal-distant energy levels described in terms of effective harmonic oscillators. In the strong interaction limit, the elastic modes decouple from the internal degrees of freedom. Since elastic modes also consists of equal-distant energy levels, we thus are led to the conclusion that the spectrum of the internal dynamics, which is that of our t-J model, also consists of equal-distant energy levels. Further work is necessary for a fully complete proof that the t-J model full spectrum takes the form Eq. (24).

Finally, we would like to point out that the states of the t-J model in the whole Hilbert space are grouped into a structure of "spin supermultiplets," as indicated by the small lattice diagonalization, similar to that of the periodic $1/r^2$ supersymmetric t-J model. Such pattern of the Hilbert space is related to the symmetries associated with the Hamiltonian. It is highly worth while to identify them more explicitly, and we would like to study these aspects in further work.

In summary, a set of Jastrow eigenfunctions have been found for the t-J model, with the eigenenergies explicitly calculated. The expected ground state of the t-J model is included in this set of wave functions. The full spectrum of the t-J model is found to have equal-distant energy levels which are highly degenerate. It would be very interesting to understand the underlying symmetry principles that give rise to such simple Hilbert space structure. It remains to study various correlation functions, as well as the thermodynamics, for this strongly correlated electron system. It would also be very interesting to study the effective field theory for the low-lying excitations for this t-J model.

Acknowledgments

We wish to thank Dr. James T. Liu and Dr. Nicolas Macris for conversations. In particular, we are very grateful to Dr. James T. Liu for his substantial numerical support. We also would like to thank the World Laboratory Foundation for the financial support.

Appendix

In this appendix, we provide a brief proof for the sum rule Eq. (9) for reader's convenience. The same argument can also be found in previous works Refs. [8, 19]. Let $X = (x_1, x_2, \ldots, x_n)$, $X_i = X/x_i$; we wish to show

$$\sum_{i=1}^{n} x_i^t \prod_{x_j \in X_i} \frac{1}{x_i - x_j} = 0, \qquad \forall t = 0, 1, 2, \ldots, (n-2). \tag{A.1}$$

Indeed, the Vandermonde determinant $V(X)$ has the property

$$\prod_{j=1}^{n} (x - x_j) = \frac{V(Xx)}{V(X)}, \tag{A.2}$$

where $X = (x_1, x_2, \ldots, x_n)$, $Xx = (x_1, x_2, \ldots, x_n, x)$. Therefore we obtain

$$\sum_{i=1}^{n} x_i^t \prod_{x_j \in X_i} \frac{1}{x_i - x_j} = \sum_{i=1}^{n} x_i^t \frac{V(X_i)}{V(X_i x_i)}$$

$$= \sum_{i=1}^{n}(-1)^{i-n}x_i^t\frac{V(X_i)}{V(X)} = \frac{(-1)^{n-1}}{V(X)}\det\begin{pmatrix} x_1^t & x_2^t & \cdots & x_n^t \\ 1 & 1 & \cdots & 1 \\ x_1 & x_2 & \cdots & x_n \\ \vdots & \vdots & \cdots & \vdots \\ x_1^{n-2} & x_2^{n-2} & \cdots & x_n^{n-2} \end{pmatrix} = 0. \tag{A.3}$$

This thus proves the sum rule Eq. (9).

Appendix B

In this appendix, we shall show that the lowering operators $A_i(z)$ and a_α give zero when acting on the Jastrow wave function ϕ given by Eq. (4). This will yield a partial confirmation that the wave function ϕ is the lowest-energy state in the subspace where the number of particles of each spin component is fixed. For the b boson degrees of freedom, we have the property

$$a_i\phi = 0, \qquad i \in (N_e+1, N_e+2, \ldots, L), \tag{B.1}$$

which is shown to be true using the sum rule Eq. (9) and the property of the Hermite polynomial roots $\sum_{j(\neq i)}(r_i - r_j)^{-1} = r_i$. The procedure to deal with the permutation operator M_{ij} in a_i is very similar to that of proving ϕ to be the eigenenergy state of the Hamiltonian, but we do not write the full details here. Combining the Eq. (B.1) with the fact $\sum_{i=1}^{L} a_i = 0$, we thus arrive at the following results:

$$\sum_{\alpha=N_e+1}^{L} a_\alpha\phi = 0,$$
$$\sum_{i=1}^{N_e} a_i\phi = 0. \tag{B.2}$$

Furthermore, we realize that

$$\left[\sum_{i=1}^{N_e} A_i(z)\right]\phi = 0. \tag{B.3}$$

We have been able to show this to be true, following the similar approach to handle the effect of the permutation operator M_{ij} acting on the Jastrow wave function ϕ. In the particular case where $\tilde{N}_\uparrow = \tilde{N}_\downarrow$, the wave function ϕ is a spin singlet and we may globally rotate Eq. (B.3) in the spin space, giving us

$$\left[\sum_{i=1}^{N_e}(A_i(\pm))\right]\phi = 0. \tag{B.4}$$

In summary, we have proved that it is impossible to construct nonvanishing eigenstates with the lowering operators and the wave functions ϕ in the subspace where the number of electrons of each flavor is fixed.

References

[1] P. W. Anderson, Science **235**, 1196 (1987).

[2] E. H. Lieb and F. Y. Wu, Phys. Rev. Lett. **20**, 1445 (1968).

[3] S. Sakar, J. Phys. A **23**, L 409 (1990); **24**, 1137 (1991); **24**, 5775 (1991); P. A. Bares, G. Blatter, and M. Ogata, Phys. Rev. B **44**, 130 (1991).

[4] B. S. Shastry, Phys. Rev. Lett. **60**, 639 (1988); **69**, 164 (1992).

[5] F. D. M. Haldane, Phys. Rev. Lett. **60**, 635 (1988); **66**, 1529 (1991).

[6] B. S. Shastry and B. Sutherland, Phys. Rev. Lett. **70**, 4092 (1993).

[7] B. Sutherland and B. S. Shastry, Phys. Rev. Lett. **75**, 5 (1993).

[8] K. Vacek, A. Okiji, and N. Kawakami (unpublished).

[9] Y. Kuramoto and H. Yokoyama, Phys. Rev. Lett. **67**, 1338 (1991).

[10] N. Kawakami, Phys. Rev. B **45**, 7525 (1992).

[11] N. Kawakami, Phys. Rev. B **46**, 1005 (1992).

[12] D. F. Wang and C. Gruber, Phys. Rev. B **49**, 15 712 (1993).

[13] D. F. Wang, James T. Liu and P. Coleman, Phys. Rev. B **46**, 6639 (1992).

[14] A. P. Polychronakos, Phys. Rev. Lett. **69**, 703 (1992); Phys. Rev. Lett. **70**, 2329 (1993).

[15] M. Fowler and J. A. Minahan, Phys. Rev. Lett. **70**, 2325 (1993).

[16] Holger Frahm, J. Phys. A **26**, 473 (1993).

[17] B. Sutherland, Phys. Rev. A **5**, 1372 (1972); **4**, 2019 (1971); J. Math. Phys. **12**, 251 (1971); **12**, 246 (1971).

[18] F. Gebhard and A. E. Ruckenstein, Phys. Rev. Lett. **68**, 244 (1992).

[19] Z. Ha and Haldane, Phys. Rev. B **46**, 9359 (1993).

[20] D. F. Wang, Q. F. Zhong and P. Coleman, Phys. Rev. B **48**, 8476 (1993).

[21] D. F. Wang, Phys. Rev. B **48**, 10 556 (1993).

QUANTUM DUALITY AND BETHE-ANSATZ FOR THE HOFSTADTER PROBLEM ON THE HEXAGONAL LATTICE

C.-A. Piguet,
D.F. Wang,
C. Gruber
Institut de Physique Théorique
Ecole Polytechnique Fédérale de Lausanne
PHB-Ecublens, CH-1015 Lausanne
Switzerland

Abstract The Hofstadter problem is studied on the hexagonal lattice. We first establish a relation between the spectra for the hexagonal lattice and for its dual lattice, the triangular lattice. Following the idea of Faddeev and Kashaev, we then obtain the Bethe-Ansatz equations for this system.

PACS: 71.30.+h; 05.30.-d; 74.65+n; 75.10.Jm

Systems in external magnetic field have been of considerable interest in recent years. One of the most fascinating properties of these systems is the integer and fractional quantum Hall effect [1, 2, 3]. The essential physics of the integer quantum Hall effect can be described by the simple Landau problem, in which a free electron moves in a two-dimensional plane under constant magnetic field. For the fractional quantum Hall effect, it is well known that the electron-electron correlation gives rise to the energy gap of the system.

Besides the systems of electrons moving continuously on the two-dimensional plane in an external magnetic field, the problem of free electrons hopping on a two-dimensional lattice under external magnetic field, i.e. the Hofstadter problem, has attracted much attention [4, 5, 6, 7, 8, 9, 10, 11]. Recently, using reflection positivity, Lieb has provided a proof for the long standing conjecture that for the square lattice, the magnetic flux which minimises the energy of the system at half-filling is exactly π per plaquette [10]. On the other hand, in their recent work, Wiegmann and Zabrodin have shown that the magnetic translations can be constructed with the generators of the quantum group $U_q(\mathrm{sl}(2))$ for the Hofstadter problem [8]. With this representation, they obtain the Bethe ansatz equations for the eigenvalue problem of the system on a two-dimensional square lattice. Later, Faddeev and Kashaev were able to provide a generalized approach, both for the square and triangular lattices [9].

In two dimensions, there are three lattices of special interest: the square, the triangular and the hexagonal lattices. In particular, the triangular lattice and the hexagonal lattice are dual of each other. In this paper, we study the Hofstadter problem on the hexagonal lattice. Following the idea of Wiegmann-Zabrodin, and Faddeev-Kashaev, we shall find the Bethe ansatz type solutions for this system. Furthermore, we establish a quantum duality relation between the energy spectra on triangular and hexagonal lattices.

The general Hamiltonian for non interacting electrons moving on a lattice in a magnetic field is

$$H = \sum_{i,j} t_{ij} \exp(i\theta_{ij}) c_i^+ c_j, \tag{1}$$

where c_i^+, c_i are the creation and annihilation operators for an electron ai site i, t_{ij} is the real hopping matrix, and θ_{ij} corresponds to $\int_i^j \boldsymbol{A}(\boldsymbol{x}) \cdot d\boldsymbol{x}$, with $\boldsymbol{A}$ the vector potential and the integral is performed on a straight line between i and j. It thus satisfies $\theta_{ij} = -\theta_{ji}$. In the following, we consider the special case where the electrons hop between nearest neighbours only. We further assume that the system is invariant under translation, but not necessarily invariant under rotation.

The Hofstadter problem is a very interesting problem of theoretical physics, in particular due to its distinction between rational and irrational numbers. In the following, we only consider the rational case when the flux per elementary cell of the lattice is given by $\phi = 2\pi M/N$, where M and N are mutually prime integers.

For the hexagonal lattice, we can define two triangular sublattices A and B. Let $\boldsymbol{s}_1$, $\boldsymbol{s}_2$ and $\boldsymbol{s}_3$ denote the three vectors that connect a site of the sublattice A with its three nearest neighbours of the sublattice B, chosen in such a way that $\boldsymbol{s}_1 \wedge \boldsymbol{s}_2$, $\boldsymbol{s}_2 \wedge \boldsymbol{s}_3$ and $\boldsymbol{s}_3 \wedge \boldsymbol{s}_1$ are in the opposite direction of the magnetic field. We have three hopping amplitudes t_1^{h}, t_2^{h} and t_3^{h} corresponding to the three directions defined by $\boldsymbol{s}_1$, $\boldsymbol{s}_2$ and $\boldsymbol{s}_3$. The index h recalls that we have a hexagonal lattice.

Let us consider a Bloch wavefunction with vector $\boldsymbol{k}$ $(0 \leq k_x, k_y \leq 2\pi/N)$,

$$|\Psi\rangle = \sum_{\boldsymbol{n}} \exp(i\boldsymbol{k} \cdot \boldsymbol{n}) u_{\boldsymbol{n}} c_{\boldsymbol{n}}^+ |0\rangle, \tag{2}$$

where the summation $\boldsymbol{n}$ is over the magnetic unit cell. We only need to consider the above region of the momentum, as in the remaining Brillouin zone the situation can be mapped to this case.

With Hamiltonian (1), the Schrödinger equation yields the following equation for the coefficients u_n corresponding to the energy E^{h},

$$E^{\mathrm{h}} u_n = \sum_{i=1}^{3} t_i^{\mathrm{h}} \alpha_i^{\mathrm{h}} \exp(i\theta_{n,n+s_i}) u_{n+s_i}, \qquad \text{if } n \in \mathrm{A} \tag{3}$$

$$E^{\mathrm{h}} u_n = \sum_{i=1}^{3} t_i^{\mathrm{h}} (\alpha_i^{\mathrm{h}})^{-1} \exp(i\theta_{n,n-s_i}) u_{n-s_i}, \qquad \text{if } n \in \mathrm{B}, \tag{4}$$

where $\alpha_i^{\mathrm{h}} = \exp(i\boldsymbol{k} \cdot \boldsymbol{s}_i)$. Combining (4) and (3), we obtain the eigenvalue equation for the coefficients u_n of the sublattice A,

$$\left\{(E^{\mathrm{h}})^2 - [(t_1^{\mathrm{h}})^2 + (t_2^{\mathrm{h}})^2 + (t_3^{\mathrm{h}})^2]\right\} u_n = \sum_{i \neq j=1}^{3} t_i^{\mathrm{h}} t_j^{\mathrm{h}} \alpha_i^{\mathrm{h}} (\alpha_j^{\mathrm{h}})^{-1} \exp(i\theta_{n,n+s_i} + i\theta_{n+s_i,n+s_i-s_j}) u_{n+s_i-s_j}. \tag{5}$$

At this point, we define the three vectors $\boldsymbol{S}_1 = \boldsymbol{s}_2 - \boldsymbol{s}_3$, $\boldsymbol{S}_2 = \boldsymbol{s}_3 - \boldsymbol{s}_1$ and $\boldsymbol{S}_3 = \boldsymbol{s}_1 - \boldsymbol{s}_2$ which connect nearest neighbours of the sublattice A. Using the fact that

$$\theta_{n,n+s_i} + \theta_{n+s_i,n+s_i-s_j} = \tfrac{1}{6}\epsilon_{ijk}\phi + \theta_{n,n+\epsilon_{ijk}S_k}, \tag{6}$$

with ϵ_{ijk} the Levi-Civita symbol and ϕ the flux through elementary hexagons, we have

$$\left\{(E^{\mathrm{h}})^2 - [(t_1^{\mathrm{h}})^2 + (t_2^{\mathrm{h}})^2 + (t_3^{\mathrm{h}})^2]\right\} u_n = \sum_{i \neq j \neq k=1}^{3} t_i^{\mathrm{h}} t_j^{\mathrm{h}} \alpha_i^{\mathrm{h}} (\alpha_j^{\mathrm{h}})^{-1} \omega^{\epsilon_{ijk}/6} \exp(i\theta_{n,n+\epsilon_{ijk}S_k}) u_{n+\epsilon_{ijk}S_k}, \tag{7}$$

where $\omega = \exp(i\phi)$.

Let us then consider the Hofstadter problem on the triangular lattice. With $t_1^{\mathrm{t}}, t_2^{\mathrm{t}}, t_3^{\mathrm{t}}$ the three hopping amplitudes the Schrödinger equation gives the following equation for the coefficients u_n corresponding to the energy E^{t},

$$E^{\mathrm{t}} u_n = \sum_{i=1}^{3} [t_i^{\mathrm{t}} \alpha_i^{\mathrm{t}} \exp(i\theta_{n,n+S_i}) u_{n+S_i} + t_i^{\mathrm{t}} (\alpha_i^{\mathrm{t}})^{-1} \exp(i\theta_{n,n-S_i}) u_{n-S_i}], \tag{8}$$

where $\alpha_i^t = \exp(i\boldsymbol{k} \cdot \boldsymbol{S}_i)$.

With the above relations, one can now easily relate the energies of the hexagonal lattice to the ones of the triangular lattice,

$$(E^{\mathrm{h}})^2 - [(t_1^{\mathrm{h}})^2 + (t_2^{\mathrm{h}})^2 + (t_3^{\mathrm{h}})^2] = E^{\mathrm{t}}, \tag{9}$$

where we have the following relations for the hopping parameters and momenta,

$$t_1^{\rm t} = t_2^{\rm h} t_3^{\rm h}, \qquad t_2^{\rm t} = t_3^{\rm h} t_1^{\rm h}, \qquad t_3^{\rm t} = t_1^{\rm h} t_2^{\rm h}$$
$$\alpha_1^{\rm t} = \alpha_2^{\rm h}(\alpha_3^{\rm h})^{-1}\omega^{1/6}, \quad \alpha_2^{\rm t} = \alpha_3^{\rm h}(\alpha_1^{\rm h})^{-1}\omega^{1/6}, \quad \alpha_3^{\rm t} = \alpha_1^{\rm h}(\alpha_2^{\rm h})^{-1}\omega^{1/6}. \tag{10}$$

It is well known that the spectrum of any bipartite lattice is symmetric around zero. This can be understood from the fact that the Hamiltonian H can be transformed in $-H$ through the following unitary transformation,

$$\begin{aligned} c_x &\longrightarrow c_x, \quad &\text{if } x \in \mathrm{A}, \\ &\longrightarrow -c_x, \quad &\text{if } x \in \mathrm{B}. \end{aligned} \tag{11}$$

The hexagonal lattice is a bipartite one, while this is not the case for the triangular lattice. Therefore the spectrum on the hexagonal lattice is symmetric around zero and this property does not hold for the triangular lattice. This result is clearly seen in Eq. (9).

The duality relation expressed by Eqs. (9) and (10) allow us to use directly the results of Faddeev and Kashaev [9] to find the Bethe ansatz equations for the hexagonal system. Let us recall briefly the results for the triangular lattice. Using a N^3 reducible representation, the Hilbert space is a tensor product of three subspaces, each of which has dimension N. A wavefunction can thus be written as

$$|\phi\rangle = |\phi\rangle_0 \otimes |\phi\rangle_1 \otimes |\phi\rangle_2. \tag{12}$$

Then, let us introduce the following notation,

$$\alpha_1^{\rm t} = e_2^{1/2}(f_2 C)^{-1/2}, \quad \alpha_2^{\rm t} = e_1^{1/2}(f_1 C)^{-1/2}, \quad \alpha_3^{\rm t} = e_0^{1/2}(f_0 C)^{-1/2},$$
$$t_1^{\rm t} = (e_2 f_2 C)^{1/2}, \qquad t_2^{\rm t} = (e_1 f_1 C)^{1/2}, \qquad t_3^{\rm t} = (e_0 f_0 C)^{1/2}. \tag{13}$$

One also has

$$e_i = b_{i-1} d_i c_{i+1}, \qquad f_i = c_{i-1} a_i b_{i+1}, \tag{14}$$

with $i \in Z_3$. Here C is a constant complex number, $C^N = (-1)^{N-1}$.

The eigenvalue equation for the projection $Q(p)$ of an eigenvector $|\phi\rangle$ on the Baxter's vector $|p\rangle$, which corresponds to the point $p = (x, \xi_0, \xi_1, \xi_2)$ on a curve in a four-dimensional space, reads

$$\Lambda(x)Q(p) = Q(\tau_- p)\Delta_-(p) + Q(\tau_+ p)\Delta_+(p), \qquad Q(p) = \langle \phi | p \rangle, \tag{15}$$

where $\Lambda(x)$ is related to the eigenvalue $E^{\rm t}$ by

$$\Lambda(x) = \omega a_0 a_1 a_2 C + d_0 d_1 d_2 + x^2 E^{\rm t}, \tag{16}$$

τ_+ and τ_- are defined on the coordinates x, ξ_0, ξ_1, ξ_2 by

$$\tau_\pm x = \omega^{\pm 1/2} x, \qquad \tau_\pm \xi_i = \omega^{-1/2} \xi_i, \qquad i = 0, 1, 2, \tag{17}$$

and

$$\Delta_-(p) = \prod_{i\in Z_3} (d_i - x\xi_{i+1}c_i), \qquad \Delta_+(p) = \prod_{i\in Z_3} \frac{\xi_i(a_i d_i - x^2 b_i c_i)}{\xi_{i+1}a_i - xb_i}. \tag{18}$$

The zeros p_k of $Q(p)$ are given by the following Bethe ansatz equation,

$$\frac{Q(\tau_- p_k)}{Q(\tau_+ p_k)} = -\frac{\Delta_+(p_k)}{\Delta_-(p_k)}. \tag{19}$$

The equations for the hexagonal lattice take the same form except that one has to introduce (10) into the definitions (13).

In the special case where the genus of the curve vanishes and N is odd ($N = 2P + 1$), one may simplify the above equations, so that they can be written out explicitly. This special case corresponds to the values of momenta: $\alpha_1^{\mathrm{t}} = \alpha_2^{\mathrm{t}} = \alpha_3^{\mathrm{t}} = q^{1/2}$, with $q = \omega^{1/2}$. The energy for the hexagonal lattice reads

$$\begin{aligned}(E^{\mathrm{h}})^2 \;=\; & (t_1^{\mathrm{h}})^2 + (t_2^{\mathrm{h}})^2 + (t_3^{\mathrm{h}})^2 + (q^{1/2} + q^{-1/2})(t_2^{\mathrm{h}} t_3^{\mathrm{h}} + t_3^{\mathrm{h}} t_1^{\mathrm{h}} + t_1^{\mathrm{h}} t_2^{\mathrm{h}}) \\ & -(q - q^{-1})[(t_1^{\mathrm{h}})^2 t_2^{\mathrm{h}} t_3^{\mathrm{h}} + t_1^{\mathrm{h}}(t_2^{\mathrm{h}})^2 t_3^{\mathrm{h}} + t_1^{\mathrm{h}} t_2^{\mathrm{h}}(t_3^{\mathrm{h}})^2] \sum_{m=1}^{2P} z_m \\ & +(t_1^{\mathrm{h}} t_2^{\mathrm{h}} t_3^{\mathrm{h}})^2 (q - q^{-1})(q^{1/2} - q^{-1/2}) \sum_{1\le m<n\le 2P} z_m z_n,\end{aligned} \tag{20}$$

where the z_l ($l = 1, \ldots, 2P$) are given by the Bethe ansatz equations,

$$\prod_{m=1,m\neq l}^{2P} \frac{qz_l - z_m}{z_l - qz_m} = q^{-1/2} \frac{(t_2^{\mathrm{h}} t_3^{\mathrm{h}} z_l + q^{1/2})(t_3^{\mathrm{h}} t_1^{\mathrm{h}} z_l + q^{1/2})(t_1^{\mathrm{h}} t_2^{\mathrm{h}} z_l + q^{1/2})}{(q^{1/2} t_2^{\mathrm{h}} t_3^{\mathrm{h}} z_l - 1)(q^{1/2} t_3^{\mathrm{h}} t_1^{\mathrm{h}} z_l - 1)(q^{1/2} t_1^{\mathrm{h}} t_2^{\mathrm{h}} z_l - 1)}. \tag{21}$$

In summary, we have developed the Bethe ansatz for the Hofstadter problem on the hexagonal lattice in this paper. An interesting duality is discovered between the hexagonal lattice and its dual partner (the triangular lattice). Using this duality relation, we have written the Bethe ansatz equations for the point of the magnetic Brillouin zone where the equations of Faddeev and Kashaev have an explicit form. Further work is necessary to obtain more information from these equations.

Acknowledgments

We wish to thank Professor H. Kunz, Professor F. Reuse, Professor S. Maumary, Professor Y.S. Wu, Professor Mo-lin Ge and Professor E.H. Lieb for conversations. This work was supported in part by the Swiss National Science Foundation.

References

[1] V. Kalmeyer and R.B. Laughlin, Phys. Rev. Lett. **59**, 2095 (1987).

[2] R. Prange and S. Girvin, eds., The Quantum Hall Effect (Springer, Berlin, 1990), and references therein.

[3] Z. Zou, B. Doucot and B.S. Shastry, Phys. Rev. B **39**, 11 424 (1989).

[4] D.J. Thouless, Phys. Rev. B **28**, 4272 (1983).

[5] P.G. Harper, Proc. Phys. Soc. A **68**, 872 (1955).

[6] J.B. Sokoloff, Phys. Rev. B **23**, 2039 (1981).

[7] D.R. Hofstadter, Phys. Rev. B **14**, 2239 (1976).

[8] P.B. Wiegmann and A.V. Zabrodin, Phys. Rev. Lett. **72**, 1890 (1994).

[9] L.D. Fadeev, R.M. Kashaev, Commun. Math. Phys. **169**, 181 (1995).

[10] E.H. Lieb, Phys. Rev. Lett. **73**, 2158 (1994); Helv. Phys. Acta **65**, 247 (1992).

[11] J. Zak, Phys. Rev. **134**, 1602 (1964).

EXACTLY SOLVABLE EXTENDED HUBBARD MODEL

D. F. Wang
Institut de Physique Théorique
Ecole Polytechnique Fédérale de Lausanne
PHB-Ecublens, CH-1015 Lausanne
Switzerland

Abstract In this work, we introduce long-range version of the extended Hubbard model. The system is defined on a nonuniform lattice. We show that the system is integrable. The ground state, the ground state energy, and the energy spectrum are also found for the system. Another long-range version of the extended Hubbard model is also introduced on a uniform lattice, and this system is proven to be integrable.

In recent years, there has been considerable interest in low-dimensional electronic systems. Systems of this type may exhibit interesting novel physics, due to low dimensionality and strong correlation. Anderson has suggested that the two-dimensional (2D) one-band Hubbard model should explain the basic physics of the high-temperature superconductivity [1]. It is suggested that the normal state of the new cuperate oxide superconductors may share the feature (Luttinger-liquid-like) of 1D interacting electron gas [1, 2]. This physical motivation has initiated lots of recent activities in one dimensional electronic models.

In one dimension, the Hubbard model was exactly solved by Lieb and Wu [3]. With the SO(4) symmetry due to Yang and Zhang [4, 5], the completeness of the eigenstates in this structure has been discussed [6]. More recently, an extended Hubbard model was introduced by several authors in general d dimensions [6]. By using η-paring mechanism, they construct eigenstates of the system, with off-diagonal-long-range order [7]. Particularly, its one-dimensional version was solved exactly with Bethe ansatz [7]. In this work, we introduce the long-range version of the extended Hubbard model in one dimension. We show that the system is completely integrable, with explicit construction of infinite constants of motion. The ground state and the energy spectrum are found for the system.

Reprinted from Wang, Phys. Rev. B 53 (1996) R1685-R1688

Let us consider a one-dimensional lattice of sites L. The positions of the sites are given by the roots of the Hermite polynomial $H_L(x)$. The roots of the Hermite polynomial $r_1, r_2, \ldots, r_L$ are all real and distinct, and this nonuniform lattice is thus well defined. Consider the following Hamiltonian:

$$H = -\frac{1}{2} \sum_{1 \le i \ne j \le L} J_{ij} \Pi_{ij}, \tag{1}$$

where $J_{ij} = 1/(r_i - r_j)^2$, and the permutation operator Π_{ij} is given by

$$\begin{aligned} \Pi_{ij} \;=\; & c^\dagger_{j\uparrow} c_{i\uparrow}(1 - n_{i\downarrow} - n_{j\downarrow}) + c^\dagger_{i\uparrow} c_{j\uparrow}(1 - n_{i\downarrow} - n_{j\downarrow}) \\ & + c^\dagger_{j\downarrow} c_{i\downarrow}(1 - n_{i\uparrow} - n_{j\uparrow}) + c^\dagger_{i\downarrow} c_{j\downarrow}(1 - n_{i\uparrow} - n_{j\uparrow}) \\ & + \tfrac{1}{2}(n_i - 1)(n_j - 1) + c^\dagger_{i\uparrow} c^\dagger_{i\downarrow} c_{j\downarrow} c_{j\uparrow} + c_{i\downarrow} c_{i\uparrow} c^\dagger_{j\uparrow} c^\dagger_{j\downarrow} \\ & - \tfrac{1}{2}(n_{i\uparrow} - n_{i\downarrow})(n_{j\uparrow} - n_{j\downarrow}) - c^\dagger_{i\downarrow} c_{i\uparrow} c^\dagger_{j\uparrow} c_{j\downarrow} - c^\dagger_{i\uparrow} c_{i\downarrow} c^\dagger_{j\downarrow} c_{j\uparrow} \\ & + (n_{i\uparrow} - 1/2)(n_{i\downarrow} - 1/2) + (n_{j\uparrow} - 1/2)(n_{j\downarrow} - 1/2), \end{aligned} \tag{2}$$

where $c_{i\sigma}$ and $c^\dagger_{i\sigma}$ are electron annihilation and creation operators at site i with spin $\sigma = \uparrow, \downarrow$. The electron number operators are given by $n_{i\uparrow} = c^\dagger_{i\uparrow} c_{i\uparrow}$, $n_{i\downarrow} = c^\dagger_{i\downarrow} c_{i\downarrow}$, $n_i = n_{i\uparrow} + n_{i\downarrow}$. If one defines the Hamiltonian on a uniform lattice and the coupling parameter $J_{ij} = \delta_{1,|i-j|}$, the system becomes the extended Hubbard model studied by Ebler, Korepin and Schoutens [7].

For the long-range extended Hubbard model defined on this nonuniform lattice, we will present its ground state, the energy spectrum and the proof of integrability in the following. At each site i, one can use the standard slave boson representation:

$$\begin{aligned} |{\downarrow\uparrow}\rangle\langle 0| = p^\dagger_i b_i, \quad & |{\downarrow\uparrow}\rangle\langle\sigma| = p^\dagger_i f_{i\sigma}, \\ |\sigma\rangle\langle\tau| = f^\dagger_{i\sigma} f_{i\tau}, \quad |\sigma\rangle\langle 0| = f^\dagger_{i\sigma} b_i, \quad & |0\rangle\langle\sigma| = b^\dagger_i f_{i\sigma}, \end{aligned} \tag{3}$$

where $|{\downarrow\uparrow}\rangle = c^\dagger_{i\downarrow} c^\dagger_{i\uparrow}|0\rangle$, $|\sigma\rangle = c^\dagger_{i\sigma}|0\rangle$, at site i. For these relations to hold, the pair operators p and the hole operators b are bosonic, while the operators f are fermionic:

$$\begin{aligned} [b_i, b^\dagger_j] &= \delta_{ij}, \\ [p_i, p^\dagger_j] &= \delta_{ij}, \\ [f_{i\sigma}, f^\dagger_{j\tau}]_+ &= \delta_{ij}\delta_{\sigma\tau}, \end{aligned} \tag{4}$$

with the constraint that at any site i there is always one particle, i.e., $\sum_{\sigma=\uparrow,\downarrow} f^\dagger_{i\sigma} f_{i\sigma} + b^\dagger_i b_i + p^\dagger_i p_i = 1$. With the slave boson representation, a state vector can be written as

$$|\phi\rangle = \sum_{\{x\},\{y\},\{z\}} \phi(x_1, x_2, \ldots, x_A | y_1\sigma_1, y_2\sigma_2, \ldots, y_M\sigma_M | z_1, z_2, \ldots, z_Q) \times \prod_{\alpha=1}^{A} p^{\dagger}_{x_\alpha} \prod_{l=1}^{M} f^{\dagger}_{y_l\sigma_l} \prod_{i=1}^{Q} b^{\dagger}_{z_i} |0\rangle, \tag{5}$$

where the wave function ϕ is symmetric when exchanging $\{x\}$ or $\{z\}$, respectively, while antisymmetric when exchanging $y_i\sigma_i$ and $y_j\sigma_j$. In the following, we use $(q_1, q_2, \ldots, q_L) = (x_1, \ldots, x_A | y_1, \ldots, y_M | z_1, \ldots, z_Q)$.

With the Hamiltonian H given by Eq. (1), the number of electron pairs, the number of holes, the number of sites single occupied by up-spin electrons, and the number of sites single occupied by the down-spin electrons are all conserved quantities. Let us denote them by the following notations:

$$A = \sum_{i=1}^{L} n_{i\downarrow} n_{i\uparrow}, \quad M_\uparrow = \sum_{i=1}^{L} n_{i\uparrow} - A, \quad M_\downarrow = \sum_{i=1}^{L} n_{i\downarrow} - A,$$
$$Q = L - M_\uparrow - M_\downarrow - A, \quad M = M_\uparrow + M_\downarrow. \tag{6}$$

In the following, we work in the Hilbert space of fixed A, $M_\uparrow$, $M_\downarrow$ and Q.

Using the amplitude ϕ defined by the Eq. (5), the eigenenergy equation can thus reduce to

$$-\frac{1}{2} \sum_{1 \le i \ne j \le L} \frac{1}{(q_i - q_j)^2} M_{ij} \phi(\{q\}) = E\phi(\{q\}), \tag{7}$$

where the operator M_{ij} permutes the coordinates q_i and q_j. The Hamiltonian $H = -\frac{1}{2}\sum_{1\le i\ne j\le L}(q_i - q_j)^{-2} M_{ij}$ commutes with the infinite number of simultaneous constants of motion, i.e., $[H, I_n] = 0$, $[I_n, I_m] = 0$, with $n, m = 1, 2, \cdots, \infty$, where $I_n = \sum_{i=1}^{L} h_i^n$, $h_i = a_i^{\dagger} a_i$, with $a_j^{\dagger} = \sum_{k(\ne j)=1}^{L} i(q_j - q_k)^{-1} M_{jk} + iq_j$, $a_i = (a^{\dagger})^{\dagger}$, with $j = 1, 2, \cdots, L$ [8, 9]. This shows that this long-range version of extended Hubbard model is also integrable. When there are no electron pairs, this system reduces to the long-range t-J model studied before [9, 10]. When there are no electron pairs and no holes, the system becomes a previous spin chain [8].

We first construct the ground-state wave function in the general subspace of fixed nonzero A, $M_\uparrow$, $M_\downarrow$ and Q. Let us look at the following Jastrow product wave function:

$$\phi(x_1, x_2, \cdots, x_A | y_1\sigma_1, y_2\sigma_2, \cdots, y_M\sigma_M | z_1, z_2, \cdots, z_Q) = \prod_{1\le i<j\le M} (y_i - y_j)^{\delta_{\sigma_i\sigma_j}} e^{i(\pi/2)\mathrm{sgn}(\sigma_i - \sigma_j)}. \tag{8}$$

This wave function is anticipated to be the ground state of the system in the subspace of fixed $A, M_\uparrow, M_\downarrow$ and Q. From previous experience [10], one may show that this wave function is indeed an eigenstate of the Hamiltonian, with the following eigenenergy:

$$E_0 = -\tfrac{1}{4}L(L-1) + \tfrac{1}{2}M_\uparrow(M_\uparrow - 1) + \tfrac{1}{2}M_\downarrow(M_\downarrow - 1). \tag{9}$$

For this Hamiltonian, one may show that $a_j^\dagger$ and a_j are raising or lowering operators defined on this lattice, $[a_j^\dagger, H] = -a_j^\dagger$, $[a_j, H] = a_j$. One may simply prove following identities:

$$\begin{aligned} a_i\phi &= 0, \qquad i = 1, 2, \ldots, M \\ a_i\phi &= 0, \qquad i = A+M+1, A+M+2, \ldots, L. \end{aligned} \tag{10}$$

Therefore, one can see that

$$\left(\sum_{i=1}^{M} a_i\right)\phi = 0, \qquad \left(\sum_{i=A+M+1}^{L} a_i\right)\phi = 0. \tag{11}$$

Furthermore, one can show that

$$\left(\sum_{i=1}^{M} a_{i+A}\sigma_i^z\right)\phi = 0. \tag{12}$$

The above three identities show the impossibilities of constructing non-zero eigenstates of lower energy than E_0 with the annihilation operators. One may regard this as partial support for our idea that ϕ is the ground state in the subspace of fixed $A, M_\uparrow, M_\downarrow, Q$. For a more complete confirmation, further work is needed in this case.

The full energy spectra of the system consists of energy levels equal-spaced,

$$E(s) = E_0 + s, \tag{13}$$

where $s = 0, 1, 2, \ldots$. For a lattice of finite size L, there is an upper bound on the value of s, as the Hilbert space of the system is finite. There are several ways of constructing excited states. The first way is to excite those electron pairs:

$$|n_1, n_2, \cdots, n_A\rangle = \sum_P \prod_{i=1}^{A} (a_i^\dagger)^{n_{P_i}} |\phi\rangle, \tag{14}$$

where $n_1, n_2, \ldots, n_A$ are integers or zero, and the summation P is over all possible permutations. Under the operation of the creation operators, any states constructed this way, if not vanishing, will be the eigenstates of the Hamiltonian with eigenenergy

$$E_1 = E_0 + (n_1 + n_2 + \cdots + n_A). \tag{15}$$

The second way to create excitations is to excite the holes of the system:

$$|m_1, m_2, \ldots, m_Q\rangle = \sum_P (a^\dagger_{A+M+1})^{m_{P_1}} (a^\dagger_{A+M+2})^{m_{P_2}} \cdots (a^\dagger_{A+M+Q})^{m_{P_Q}} |\phi\rangle, \tag{16}$$

with the eigenenergies given by

$$E_2 = E_0 + (m_1 + m_2 + \cdots + m_Q), \tag{17}$$

if these state vectors are not zero. Finally, one may create excitations by exciting the electrons on the single-occupied sites. These excited states may be written as

$$|\phi_3\rangle = \sum_P [S_1(\nu_1)^\dagger]^{n_{P_1}} [S_2(\nu_2)^\dagger]^{n_{P_2}} \cdots [S_M(\nu_M)^\dagger]^{n_{P_M}} |\phi\rangle, \tag{18}$$

where $S_i(\nu_i)^\dagger = a^\dagger_{i+A}\sigma_i^{\nu_i}$, with $\nu_i = 0, \pm, z$. These states have eigenenergies given by

$$E_3 = E_0 + (n_1 + n_2 + \cdots + n_M). \tag{19}$$

Clearly, excited energy levels are highly degenerate. However, we are still unable to develop a systematic rule to characterize the pattern of the degeneracy and to explain it with symmetries of the system (presumably, Yangian symmetry).

In the limiting case where $M = M_\uparrow + M_\downarrow = 0$, one has two types of bosons on the chain, the local electron pairs and the holes. The wave function ϕ becomes

$$\phi(x_1, x_2, \ldots, x_A | z_1, z_2, \ldots, z_Q) = 1, \tag{20}$$

which is obviously the ground state of the Hamiltonian H, with the ground state energy $E_0 = -L(L-1)/4$. With the raising operators, one may construct the first excited state quite easily:

$$\phi'(x_1, x_2, \ldots, x_A | z_1, z_2, \ldots, z_Q) = (x_1 + x_2 + \cdots + x_A). \tag{21}$$

The wave functions of higher excited states may be obtained by acting on this wave function again with raising operators.

Besides this extended Hubbard model of long-range type defined on a nonuniform lattice, we would like to introduce another model of the extended Hubbard model of long-range type. Consider a uniform lattice of sites equally spaced. The size of the lattice is L. The Hamiltonian of the system is given by

$$H = -\frac{1}{2} \sum_{1 \le i \ne j \le L} d(i-j)^{-2} \Pi_{ij}, \tag{22}$$

where the function $d(i-j) = (L/\pi)\sin[\pi(i-j)/L]$, and the permutation operator Π_{ij} is the same as Eq. (2). From our previous experience, one can simply write the first quantized Hamiltonian as follows:

$$H = -\frac{1}{2} \sum_{1 \le i \ne j \le L} 1/d^2(q_i - q_j) M_{ij}. \tag{23}$$

It is very easy to prove that this extended Hubbard model is also integrable, once the Hamiltonian is written this way by using the relation of Fowler and Minahan [11]:

$$L_n = \sum_{i=1}^{L} \pi_i^n, \quad \pi_i = \sum_{j(\ne i)=1}^{L} z_j/(z_i - z_j) M_{ij},$$
$$[L_n, L_m] = 0, \quad [L_n, H] = 0, \quad n, m = 1, 2, \ldots, \tag{24}$$

where $z_j = e^{2\pi i q_j/L}$, $j = 1, 2, \ldots, L$. This thus provides a proof for the integrability of the extended Hubbard model. In the limiting case of no electron pairs, this system reduces to the long-range t-J model [12, 13, 14, 9]. If there are no holes and no electron pairs, the Hamiltonian reduces to the Haldane-Shastry spin chain [15, 16]. In the general case, to write the conserved quantities of the system in terms of second quantization, one may simply use the permutation symmetries of the wave function. In this paper, we shall not present the wave functions and the eigenenergy spectrum of this system. More results on this system will be given elsewhere.

In summary, we have introduced two integrable models of strongly correlated electrons, one defined on a nonuniform lattice, the other on a uniform lattice. The systems belong to the Jastrow-integrable type. Particularly, the system on the nonuniform lattice has an equal-spacing energy spectrum. Because of the degrees of freedom of the paired electrons, the energy levels will have much larger degeneracy than the long-range t-J model. One of the most interesting things is to determine explicitly the underlying symmetries that account for the degeneracy of each energy level.

Acknowledgments

I wish to thank V. E. Korepin for interesting conversations. Discussions with Mo-lin Ge and C. Gruber are acknowledged gratefully. This work was supported by the Swiss National Science Foundation.

References

[1] P. W. Anderson, Science **235**, 1196 (1987).

[2] Lu Yu, Zhao-bin Su and Yan-min Li, Chin. J. Phys. (Taipei) **31**, 579 (1993), and references therein.

[3] E. H. Lieb and F. Y. Wu, Phys. Rev. Lett. **20**, 1445 (1968), C. N. Yang, *ibid.* **19**, 1312 (1967).

[4] C. N. Yang, Phys. Rev. Lett. **63**, 2144 (1989).

[5] C. N. Yang and S. C. Zhang, Mod. Phys. Lett. B **4**, 759 (1990).

[6] F. H. L. Ebler, V. E. Korepin and K. Schoutens, Nucl. Phys. B **384**, 431 (1992).

[7] F. H. L. Ebler, V. E. Korepin and K. Schoutens, Phys. Rev. Lett. **68**, 2960 (1993).

[8] A. P. Polychronakos, Phys. Rev. Lett. **70**, 2329 (1993).

[9] D. F. Wang and C. Gruber, Phys. Rev. B **49**, 15 712 (1994).

[10] C. Gruber and D. F. Wang, Phys. Rev. B **50**, 3103 (1994).

[11] M. Fowler and J. A. Minahan, Phys. Rev. Lett. **70**, 2325 (1993).

[12] Y. Kuramoto and H. Yokoyama, Phys. Rev. Lett. **67**, 1338 (1991).

[13] N. Kawakami, Phys. Rev. B **46**, 1005 (1992).

[14] D. F. Wang, James T. Liu and P. Coleman, Phys. Rev. B **46**, 6639 (1992).

[15] F. D. M. Haldane, Phys. Rev. Lett. **60**, 635 (1988).

[16] B. S. Shastry, Phys. Rev. Lett. **60**, 639 (1988).

References

INTEGRABILITIES OF THE LONG-RANGE t-J MODELS WITH TWISTED BOUNDARY CONDITIONS

James T. Liu
Department of Physics
The Rockefeller University
1230 York Avenue
New York, NY 10021-6399

D. F. Wang
Institut de Physique Théorique
Ecole Polytechnique Fédérale de Lausanne
PHB-Ecublens, CH-1015 Lausanne
Switzerland

Abstract The integrability of the one-dimensional long-range supersymmetric t-J model has previously been established for both open systems and those closed by periodic boundary conditions through explicit construction of its integrals of motion. Recently the system has been extended to include the effect of magnetic flux, which gives rise to a closed chain with twisted boundary conditions. While the t-J model with twisted boundary conditions has been solved for the ground state and full energy spectrum, proof of its integrability has so far been lacking. In this letter we extend the proof of integrability of the long-range supersymmetric t-J model and its SU($m|n$) generalization to include the case of twisted boundary conditions.

Solvable models have attracted attention from both the high-energy and condensed-matter communities. These models provide important examples where it is possible to deal with many degrees of freedom without having to resort to perturbation theory. Interesting models in condensed-matter physics that have been solved include the short-range spin model [1], delta-function bose gas [2], delta-function electron gas [3, 4], the Hubbard model [5], the Luttinger model [6], the magnetic impurity model and the Anderson impurity model [7, 8].

Reprinted from Liu *et al.*, Phys. Rev. B 55 (1997) R3344-R3346

In one dimension, ever since Haldane and Shastry independently introduced the $1/r^2$ spin model [9, 10], there has been considerable interest in the model and its generalizations, such as the long range supersymmetric t-J model [11, 12, 13, 14] and its multicomponent SU($m|n$) version (where m and n denote the number of bosonic and fermionic species, respectively). All of these models are characterized by having a ground-state wave function which takes on a Jastrow product form, and by having quasiparticle scattering matrices of a very simple form, as in the continuous Calogero-Sutherland systems describing nonrelativistic quantum particles [15]. In particular, the Haldane-Shastry spin model can also be identified as a free system composed of identical particles obeying Haldane's generalized Pauli principle [16], and obeying a generalized statistical distribution function at finite temperature [17]. In 1992, Gebhard and Ruckenstein introduced the long-range Hubbard model, in which the electrons are described by the $1/r$ Hubbard model. It is noteworthy that this $1/r$ Hubbard model is integrable for any on-site energy; the full energy spectrum and thermodynamics have been solved explicitly [18]. At half-filling and in the limit of large interaction, this model reduces to the SU(2) Haldane-Shastry spin chain. For less than half-filling, but still in the limit of $U = \infty$, the system remains characterized by eigenfunctions of a Jastrow product form.

Recently there has been considerable interest in adding magnetic flux to the Haldane-Shastry-type models. For a one-dimensional ring threaded by flux, this reduces to the problem of incorporating twisted boundary conditions. A twisted version of the long-range integrable Haldane-Shastry spin chain has been introduced, and was solved in the rational flux case [19]. Subsequently, this was generalized to the case of the long-range t-J model with twisted boundary conditions [20, 21]. In particular, it was shown that the irrational flux case can be treated identically [21], indicating that there is no essential difference between rational and irrational flux. Based on the exact solutions, it is natural to expect that the long-range models remain integrable despite the twisted boundary conditions. However until now this has remained an open problem. In this letter, we provide a proof of the integrability of the long-range t-J model and its SU($m|n$) generalization with twisted boundary conditions by explicitly constructing an infinite number of simultaneous constants of motion. This construction is a straightforward extension of the methods used in the absence of flux [22, 23, 24, 25, 26, 27], and is motivated by the mapping of the closed ring onto an equivalent open system where the flux is manifested in twisted boundary conditions. A further consequence of this mapping is that it yields a unified treatment of the integrability of both the open and closed chains.

Because of the subtleties involved with introducing magnetic flux into a model with long-range interactions, we follow the procedure of [19], and start with an open chain which is subsequently closed through appropriate boundary conditions to form a ring of N sites. The Hamiltonian of the SU(1|2) t-J model on this open lattice takes the form

$$H_0 = \frac{1}{2}P_G \sum_{\alpha\neq\beta} \frac{1}{(q_\alpha - q_\beta)^2}\Bigg[-\sum_\sigma (c^\dagger_{\alpha\sigma}c_{\beta\sigma} + \text{H.c.}) + \hat{P}_{\alpha,\beta} - (1-n_\alpha)(1-n_\beta)\Bigg] P_G, \tag{1}$$

where P_G is the Gutzwiller projection onto single occupancy and the operators c and $c^\dagger$ are the usual fermion operators that satisfy standard fermionic anticommutation relations. The summation over σ is to sum over all the spin components of the electrons, $\sigma = \uparrow, \downarrow$ for SU(1|2). The spin exchange operator $\hat{P}_{\alpha,\beta}$ is given by

$$\hat{P}_{\alpha,\beta} = \sum_\sigma \sum_{\sigma'} c^\dagger_{\alpha\sigma}c_{\alpha\sigma'}c^\dagger_{\beta\sigma'}c_{\beta\sigma}, \tag{2}$$

and the electron density operator is $n_\alpha = \sum_\sigma c^\dagger_{\alpha\sigma}c_{\alpha\sigma}$. The lattice permutation form of this Hamiltonian may be made explicit by introducing the *graded* permutation operator $\Pi^{\nu,\nu'}_{\alpha,\beta}$, which exchanges particles of species ν and ν' at locations α and β (where $\nu, \nu' = 0, \uparrow, \downarrow$ with 0 denoting a hole). Written in terms of Π, the Hamiltonian becomes

$$H_0 = -\frac{1}{2}\sum_{\alpha\neq\beta}\sum_{\nu,\nu'} \frac{\Pi^{\nu,\nu'}_{\alpha,\beta}}{(q_\alpha - q_\beta)^2}, \tag{3}$$

where the Gutzwiller projection is no longer necessary since the graded permutation operator does not generate doubly occupied states. In this form, the SU($m|n$) generalization is immediately obvious.

For a finite open chain with L sites, integrability is achieved when the lattice positions q_α lie at the roots of the Lth Hermite polynomial [22]. In the limit $L \to \infty$ these roots become equally spaced, and translational invariance is restored. It is precisely in this limit that it is possible to close the chain by demanding twisted boundary conditions. We allow a separate twist ϕ_ν for each independent species ν, so that the twisted boundary conditions for a ring of N sites may be encoded by the requirement that

$$\Pi^{\nu,\nu'}_{\alpha,\beta+lN} = z^{lN(\phi_\nu - \phi_{\nu'})}\Pi^{\nu,\nu'}_{\alpha,\beta}, \tag{4}$$

where z is the primitive Nth root of unity. Note that the basic period for the spectral flow corresponds to ϕ_ν running from 0 to 1. Since the resulting closed system is translationally invariant, we single out one period and reexpress the Hamiltonian, after gauge transformation, as $H_0 = E_0 + H$ where [19]

$$H = -\frac{1}{2}\sum_{1\leq i\neq j\leq N}\sum_{\nu,\nu'}\sum_{l=-\infty}^{\infty} \frac{z^{(i-j-lN)(\phi_\nu - \phi_{\nu'})}\Pi^{\nu,\nu'}_{i,j}}{(i-j-lN)^2}$$

$$= -\frac{1}{2} \sum_{1 \le i \ne j \le N} \sum_{\nu,\nu'} J_{\phi_\nu - \phi_{\nu'}}(i-j) \Pi_{i,j}^{\nu,\nu'} \tag{5}$$

now defines the long-range supersymmetric t-J model on a periodic ring. The offset $E_0 = \pi^2/3N$ accounts for exchanges lN units apart (which is present in H_0 but not in the periodic H), and may be interpreted as a shift in the ground-state energy from finite-size effects. The sum over l ensures the appropriate periodicity of the ring under translations, and yields the inverse trigonometric potential [21]

$$\begin{aligned} J_\phi(n) &= \sum_l \frac{z^{(n+lN)\phi}}{(n+lN)^2} \\ &= \left(\frac{\pi}{N}\right)^2 \frac{z^{\lfloor\phi\rfloor n}[1 + (\phi - \lfloor\phi\rfloor)(z^n - 1)]}{\sin^2(n\pi/N)}. \end{aligned} \tag{6}$$

This expression is piecewise linear and continuous in ϕ, leading to many remarkable features of this model [19, 20, 21]. In the case of periodic boundary conditions ($\phi = 0$), the physical properties of the supersymmetric t-J model on a uniform closed chain have been studied previously [11, 12, 13, 14].

To provide a proof for the long-range t-J model with twisted boundary conditions, we are motivated by the previous results on the integrabilities of the uniform long-range t-J model with periodic boundary conditions, and the non-uniform long-range t-J model with open boundary conditions [26, 27], which were generalizations of the spin chain results [22, 23, 24, 25]. Proof of the integrability proceeds by first mapping the species-exchange Hamiltonian, Eq. (5), to a lattice permutation equivalent using slave-boson techniques. The resulting Hamiltonian then acts on wave functions ψ written in the form $\psi(q_1\nu_1, q_2\nu_2, \ldots, q_N\nu_N)$, where q_i and ν_i label the position and SU($m|n$) "spin" of particle i. Acting on such wave functions, and using the fact that $\{q_i\}$ span the lattice due to single occupancy, the Hamiltonian becomes

$$H = -\frac{1}{2} \sum_{1 \le i \ne j \le N} J_{\phi_{\nu_i} - \phi_{\nu_j}}(q_i - q_j) M_{ij}, \tag{7}$$

where the particle exchange operator M_{ij} is defined by

$$M_{ij}\psi(\ldots, q_i\sigma_i, \ldots, q_j\sigma_j, \ldots) \equiv \psi(\ldots, q_j\sigma_i, \ldots, q_i\sigma_j, \ldots). \tag{8}$$

Note that the fermionic and bosonic nature of the individual species is fully encoded in the wavefunctions; $\psi \to \pm\psi$ under simultaneous interchange of position and spin. This independence of the exchange operator from the particle statistics ensures that the proof of integrability holds for *all* SU($m|n$) extended t-J models, and not just for the SU(1|2) case.

Based on the integrability proof for the open chain and for the ring closed by periodic boundary conditions studied in Refs. [22] and [23], we introduce the generalized operators

$$\pi_j = i \sum_{k \neq j} u_{\phi_{\nu_j} - \phi_{\nu_k}} (q_j - q_k) M_{jk}, \tag{9}$$

where $u_\phi(n)$ is the (twisted) periodic version of $1/r$

$$u_\phi(n) = \sum_l \frac{z^{(n+lN)\phi}}{n + lN}. \tag{10}$$

As in the case for $J_\phi(n)$, this sum may be performed, yielding

$$u_\phi(n) = \frac{2\pi i}{N} \frac{z^{\lfloor \phi \rfloor n}}{1 - z^{-n}} \tag{11}$$

(for nonintegral ϕ). Note that ϕ enters discontinuously, with a jump in $u_\phi(n)$ at integral values of ϕ. A careful treatment of convergence issues for integral ϕ indicates that the actual value of the infinite sum in Eq. (10) is the average of the values of $u_\phi(n)$ before and after the discontinuity. Nevertheless, for a consistent treatment of the invariants, we take Eq. (11) as the definition of $u_\phi(n)$ for *all* values of ϕ. A consequence of this asymmetry is to pick a preferred ordering, thus breaking the parity symmetry $u_{-\phi}(-n) = -u_\phi(n)$, which otherwise holds for nonintegral ϕ. Nevertheless, this particular choice of ordering gives

$$u_0(n) = \frac{2\pi i}{N} \frac{1}{1 - z^{-n}}, \tag{12}$$

in agreement with previous results in the absence of flux [22, 23]. Overall, this subtle treatment of integral twists indicates that, surprisingly enough, it is actually the *zero* flux case that is exceptional; this case corresponds to working on top of the locations of the cusps in the spectral flow itself.

Explicit evaluation of the commutators brings out a distinction between integral and non-integral twists:

$$[\pi_i, \pi_j] = \begin{cases} \frac{2\pi}{N} z^{\phi_{ij} q_{ij}} M_{ij} (\pi_i - \pi_j) & \text{for } \phi_{ij} \in Z \\ 0 & \text{otherwise,} \end{cases} \tag{13}$$

where $\phi_{ij} = \phi_{\nu_i} - \phi_{\nu_j}$ and $q_{ij} = q_i - q_j$. Using the relation $(z^{\phi_{ij} q_{ij}} M_{ij}) \pi_j = \pi_i (z^{\phi_{ij} q_{ij}} M_{ij})$, valid whenever $\phi_{ij} \in Z$, the above commutator may be reexpressed in a form similar to that of Ref. [23],

$$[\pi_i, \pi_j] = \begin{cases} \frac{2\pi}{N} [z^{\phi_{ij} q_{ij}} M_{ij}, \pi_i] & \text{for } \phi_{ij} \in Z \\ 0 & \text{otherwise.} \end{cases} \tag{14}$$

From here it is obvious that the commutation results of Fowler and Minahan [23] are easily generalized to the present case. Therefore the infinite set of hermitian operators

$$I_M = \sum_i \pi_i^M \tag{15}$$

(where $M = 0, 1, 2, \ldots$), provides a set of mutually commuting operators, $[I_M, I_N] = 0$, regardless of the individual species twists. Note that the commutation is trivial for nonintegral relative twists, and is basically a consequence of the simple open chain result, $[\pi_{0\,\alpha}, \pi_{0\,\beta}] = 0$, where

$$\pi_{0\,\alpha} = i \sum_{\beta \neq \alpha} \frac{1}{q_\alpha - q_\beta} M_{\alpha\beta} \tag{16}$$

is the corresponding open chain operator [compare with Eq. (9)].

For a finite open chain, it is known that the π_0 operators do not commute with the Hamiltonian [22],

$$[H_0, \pi_{0\,\alpha}] = 2i \sum_{\beta \neq \alpha} \frac{1}{(q_\beta - q_\alpha)^3}. \tag{17}$$

However, in making the system periodic, the odd exponent in Eq. (17) allows the cancellation of terms exchanging to the left and right. As a result, we find

$$[H, \pi_i] = 0 \tag{18}$$

for the periodic case, regardless of the relative twist angles. It is now clear that the mutually commuting set of operators, $\{I_M\}$ for $M = 0, 1, 2, \ldots$, provide explicit constants of motion of the Hamiltonian H, and hence proves the integrability of the long range SU$(m|n)$ t-J model on a ring (for either twisted or untwisted boundary conditions).

In conclusion, we have provided a proof for the integrability of the long-range t-J models with twisted boundary conditions by explicitly constructing an infinite set of mutually commuting constants of motion. This proof generalizes previous results for rings without flux, and makes use of the viewpoint that the closed chain is simply a periodic version of the open system. A consequence of this similar treatment for both closed and open chains is the demonstration that the key property behind the integrability of these models is simply the permutation nature of the system. These results have filled a gap in that the integrability condition for the twisted t-J model was as yet unknown, in spite of the fact that several thorough studies of the long-range model in the presence of flux have been provided.

Acknowledgments

We wish to thank C. Gruber, H. Kunz, R. Khuri and H. C. Ren for stimulating discussions. In addition, one of us (D.F.W.) wishes to thank P. Coleman for encouragement. This work was supported in part by the U.S. Department of Energy under Grant No. DOE-91ER40651-TASKB, and by the Swiss National Science Foundation.

References

[1] H. Bethe, Z. Phys. **75**, 205 (1931).

[2] E. H. Lieb and Liniger, Phys. Rev. **130**, 1605 (1963); E. H. Lieb, *ibid.* **130**, 1616 (1963).

[3] F. Flicker and E. H. Lieb, Phys. Rev. **161**, 179 (1967).

[4] C. N. Yang, Phys. Rev. Lett. **19**, 1312 (1967); M. Gaudin, Phys. Lett. A **24**, 55 (1967).

[5] E. H. Lieb and F. Y. Wu, Phys. Rev. Lett. **20**, 1445 (1968); C. N. Yang, *ibid.* **63**, 2144 (1989); C. N. Yang and S. C. Zhang, Mod. Phys. Lett. B **4**, 759 (1990).

[6] D. C. Mattis and E. H. Lieb, J. Math. Phys. (N.Y.) **6**, 304 (1965).

[7] N. Andrei, K. Furuya and J. H. Lowenstein, Rev. Mod. Phys. **55**, 331 (1983), and references therein.

[8] P. B. Wiegmann, J. Phys. A **14**, 1463 (1981).

[9] F. D. M. Haldane, Phys. Rev. Lett. **60**, 635 (1988).

[10] B. S. Shastry, Phys. Rev. Lett. **60**, 639 (1988).

[11] Y. Kuramoto and H. Yokoyama, Phys. Rev. Lett. **67**, 1338 (1991).

[12] N. Kawakami, Phys. Rev. B **46**, 1005 (1992).

[13] D. F. Wang, J. T. Liu and P. Coleman, Phys. Rev. B **46**, 6639 (1992).

[14] Z. N. C. Ha and F. D. M. Haldane, Phys. Rev. B **46**, 9359 (1992).

[15] B. Sutherland J. Math. Phys. (N.Y.) **12**, 246 (1971); **12**, 251 (1971); Phys. Rev. A **4**, 2019 (1971); **5**, 1372 (1972); F. Calogero, J. Math. Phys. (N.Y.) **10**, 2191 (1969); **10**, 2197 (1969).

[16] F. D. M. Haldane, Phys. Rev. Lett. **66**, 1529 (1991).

[17] Y. S. Wu, Phys. Rev. Lett. **73**, 922 (1994).

[18] F. Gebhard and A. Ruckenstein, Phys. Rev. Lett. **68**, 244 (1992).

[19] T. Fukui and N. Kawakami, Phys. Rev. Lett. **76**, 4242 (1996).

[20] T. Fukui and N. Kawakami, Phys. Rev. B **54**, 5346 (1996).

[21] James T. Liu and D. F. Wang (unpublished).

[22] A. P. Polychronakos, Phys. Rev. Lett. **69**, 703 (1992); **70**, 2329 (1993).

[23] M. Fowler and J. A. Minahan, Phys. Rev. Lett. **70**, 2325 (1993).

[24] L. Brink, T. H. Hansson and M. A. Vasiliev, Phys. Lett. B **286**, 109 (1992).

[25] V. I. Inozemtsev, J. Stat. Phys. **59**, 1143 (1990).

[26] D. F. Wang and C. Gruber, Phys. Rev. B **49**, 15 712 (1994).

[27] C. Gruber and D. F. Wang, Phys. Rev. B **50**, 3103 (1994).

SU($m|n$) SUPERSYMMETRIC CALOGERO-SUTHERLAND MODEL CONFINED IN HARMONIC POTENTIAL

C.-A. Piguet,
D. F. Wang
and C. Gruber
Institut de Physique Théorique
Ecole Polytechnique Fédérale de Lausanne
PHB-Ecublens, CH-1015 Lausanne
Switzerland

Abstract In this work, we study a continuous quantum system of a mixture of bosons and fermions with the supersymmetry SU($m|n$). The particles are confined in a harmonic well and interact with each other through the $1/r^2$ interaction. An eigenstate of the Hamiltonian is constructed explicitly for the most general SU($m|n$) case and its energy is given explicitly. It is argued that this eigenstate is the ground state of the system. Moreover, we construct some excited states giving an equally spaced spectrum. In the limiting case where there are no bosons in the system, our results reduce to those obtained previously.

There have been considerable recent interests in study of the Calogero-Sutherland models [1, 2], as well as their various generalizations. One of the examples is the multi-component generalization of the CS model. For the SU(n) fermionic particles confined in a harmonic oscillator potential and interacting with each other through the $1/r^2$ interaction, the ground state wavefunction was constructed explicitly, and the ground state energy was found in previous work [6]. The excitation of the fermionic system was found to be equal spaced [3, 4, 5, 6, 8]. On the other hand, the system consisting of bosons with SU(n) spins has been studied for the pairwise interaction $\lambda(\lambda + P_{ij}^{\sigma})/x_{ij}^2$ in presence of harmonic potential confinement, where P_{ij}^{σ} permutes the spins of the bosons [6].

In the following, we will investigate the ground state wavefunction of the the most general supersymmetric SU($m|n$) multi-component CS model confined in a harmonic potential. In this case, the system consists of a mixture of bosons and fermions. The ground state wavefunction for this supersymmetric case has

Reprinted from Piguet *et al.*, Int. J. Mod. Phys. B 11 (1997) 1839-1844 been unknown so far. In this work, we provide the ground state wavefunction, and compute its energy explicitly in the subspace where the number of particles of each flavor is fixed. In the limiting case where there are only fermions in the system, our results reduce to those obtained by previous authors.

We consider a one-dimensional quantum system of Q interacting particles confined in a harmonic potential of frequency ω. The Hamiltonian under consideration reads

$$H = -\frac{1}{2}\sum_{i=1}^{Q} \frac{\partial^2}{\partial q_i^2} + \frac{\omega^2}{2}\sum_{i=1}^{Q} q_i^2 + \sum_{j>i} \frac{\lambda(\lambda - Y_{ij})}{(q_i - q_j)^2} \tag{1}$$

where $q_1, \ldots, q_Q$ are the coordinates of the particles and Y_{ij} is the operator that exchanges the positions of particles i and j. The real constant λ is supposed to be greater than $1/2$. This Hamiltonian is exactly the one considered in Ref. [6] for a system of fermions only.

In this work, we consider the case where we have M fermions and N bosons ($M + N = Q$) with SU$(m|n)$ spin degrees of freedom. We denote $x_1, \ldots, x_M$ the positions of the fermions, $y_1, \ldots, y_N$ the positions of the bosons, $\sigma_1, \ldots, \sigma_M$ ($\sigma_i \in \{1, \ldots, m\}$) the spins of the fermions and $\tau_1, \ldots, \tau_N$ ($\tau_\alpha \in \{1, \ldots, n\}$) the spins of the bosons. We also use the notation that $(q_1, q_2, \ldots, q_Q) = (x_1, \ldots, x_M | y_1, y_2, \ldots, y_N)$. With these notations, the Hamiltonian Eq. (1) is made of three parts

$$H = H_F + H_B + H_M \tag{2}$$

where

$$\begin{aligned} H_F &= -\frac{1}{2}\sum_{i=1}^{M} \frac{\partial^2}{\partial x_i^2} + \frac{\omega^2}{2}\sum_{i=1}^{M} x_i^2 + \sum_{j>i} \frac{\lambda(\lambda - Y_{ij})}{(x_i - x_j)^2} \\ H_B &= -\frac{1}{2}\sum_{\alpha=1}^{N} \frac{\partial^2}{\partial y_\alpha^2} + \frac{\omega^2}{2}\sum_{\alpha=1}^{N} y_\alpha^2 + \sum_{\beta>\alpha} \frac{\lambda(\lambda - Y_{\alpha\beta})}{(y_\alpha - y_\beta)^2} \\ H_M &= \sum_{i,\alpha} \frac{\lambda(\lambda - Y_{i\alpha})}{(x_i - y_\alpha)^2}. \end{aligned} \tag{3}$$

H_F (H_B) is the Hamiltonian for a system of M (N) interacting fermions (bosons). H_M represents the interaction between the fermions and the bosons.

We now wish to find the ground state wavefunction of the Hamiltonian. For this, we propose a wavefunction which is a product of three terms

$$\begin{aligned} &\Psi(x_1, \sigma_1; \ldots; x_M, \sigma_M | y_1, \tau_1; \ldots; y_N, \tau_N) \\ &\quad = \Psi_F(x_1, \sigma_1, \ldots, x_M, \sigma_M) \cdot \Psi_B(y_1, \ldots, y_N) \\ &\quad \cdot \Psi_M(x_1, \ldots, x_M, y_1, \ldots, y_N) \end{aligned} \tag{4}$$

where

$$\begin{aligned}\Psi_F(x_1,\sigma_1;\ldots;x_M,\sigma_M) &= \prod_{j>i}|x_j-x_i|^\lambda(x_j-x_i)^{\delta_{\sigma_i\sigma_j}}\\ &\quad\times\exp\left(i\frac{\pi}{2}\text{sgn}(\sigma_i-\sigma_j)\right)\prod_i\exp\left(-\frac{\omega}{2}x_i^2\right)\\ \Psi_B(y_1,\ldots,y_N) &= \prod_{\beta>\alpha}|y_\beta-y_\alpha|^\lambda\prod_\alpha\exp\left(-\frac{\omega}{2}y_\alpha^2\right)\\ \Psi_M(x_1,\ldots,x_M,y_1,\ldots,y_N) &= \prod_{i,\alpha}|x_i-y_\alpha|^\lambda.\end{aligned} \tag{5}$$

Let us first remark that this wavefunction has the correct symmetry: $M_{ij}\Psi = -\Psi$ for the fermions and $M_{\alpha\beta}\Psi = \Psi$ for the bosons, where M is the operator that exchanges two particles. We can also remark that the spin of the bosons does not enter this wavefunction.

Let us now show that the wavefunction Eq. (4) is an eigenstate of the Hamiltonian. The wavefunction Ψ_F is the ground state of the system with only fermions found in Ref. [6]. We thus have

$$H_F\Psi_F = \frac{\omega}{2}\left[\lambda M(M-1)+\sum_{k=1}^{m}M_k^2\right]\Psi_F \tag{6}$$

where M_k is the number of fermions with spin k. The wavefunction Ψ_B is the ground state of the system with only bosons. It is easy to check that

$$H_B\Psi_B = \frac{\omega}{2}\left[\lambda N(N-1)+N\right]\Psi_B. \tag{7}$$

With this, one can show that

$$\begin{aligned}\frac{1}{\Psi}(H_F+H_B)\Psi = \frac{\omega}{2}&\left[\lambda\{M(M-1)+N(N-1)+2MN\}+\sum_{k=1}^{m}M_k^2+N\right]\\ &-\sum_{i,\alpha}\frac{\lambda(\lambda-1)}{(x_i-y_\alpha)^2}-\sum_{\alpha,i\neq j}\frac{\lambda\delta_{\sigma_i\sigma_j}}{(x_i-x_j)(x_i-y_\alpha)}.\end{aligned} \tag{8}$$

To apply the interaction part on the wavefunction, we first compute the effect of $Y_{i\alpha}$

$$\frac{Y_{i\alpha}\Psi}{\Psi} = \prod_{j\neq i}\left(\frac{x_j-y_\alpha}{x_j-x_i}\right)^{\delta_{\sigma_i\sigma_j}} = \prod_{j\neq i}\left(1+\delta_{\sigma_i\sigma_j}\frac{x_i-y_\alpha}{x_j-x_i}\right). \tag{9}$$

Thus

$$\frac{1}{\Psi}\left[\sum_{i,\alpha}\frac{\lambda(\lambda-K_{i\alpha})}{(x_i-y_\alpha)^2}\Psi\right]$$

$$= \sum_{i,\alpha} \frac{\lambda^2}{(x_i - y_\alpha)^2} - \sum_{i,\alpha} \frac{\lambda}{(x_i - y_\alpha)^2} \prod_{j \neq i} \left(1 + \delta_{\sigma_i \sigma_j} \frac{x_i - y_\alpha}{x_j - x_i} \right) \tag{10}$$

We can expand the second term in power of $x_i - y_\alpha$. The terms of order greater than 1 are zero as will be proved in the following. Using this result, we have

$$\frac{1}{\Psi} \left[\sum_{i,\alpha} \frac{\lambda(\lambda - K_{i\alpha})}{(x_i - y_\alpha)^2} \Psi \right] = \sum_{i,\alpha} \frac{\lambda(\lambda - 1)}{(x_i - y_\alpha)^2} - \sum_{\alpha, i \neq j} \frac{\lambda \delta_{\sigma_i \sigma_j}}{(x_i - y_\alpha)(x_j - x_i)}. \tag{11}$$

These terms cancel with the fermionic and bosonic nonconstant terms in Eq. (8) to give the following ground state energy:

$$E_G = \frac{\omega}{2} \left\{ \lambda[M(M-1) + N(N-1) + 2MN] + \sum_{k=1}^{m} M_k^2 + N \right\}. \tag{12}$$

We have now to prove that the terms of order greater than 1 of the second term of Eq. (10) are zero. For this, let us consider a term of order s. We have $s+1$ particles with the same spin with coordinates $i, k_1, \ldots, k_s$. Let us now show that if we sum over all the permutations $i \leftrightarrow k_j$ with $j = 1, \ldots, s$ while keeping α fixed we get a zero contribution:

$$-\lambda \sum_{\{i,k_1,\ldots,k_s\}} (x_i - y_\alpha)^{s-2} \prod_{j=i}^{s} \frac{1}{x_{k_j} - x_i} = 0. \tag{13}$$

For this, following Ref. [6], we expand the term $(x_i - y_\alpha)^{s-2}$ and express the product in terms of Vandermonde determinant

$$\prod_{j=i}^{s} \frac{1}{x_{k_j} - x_i} = (-1)^s \frac{V^{(s)}(x_{k_1}, \ldots, x_{k_s})}{V^{(s+1)}(x_{k_1}, \ldots, x_{k_s}, x_i)} \tag{14}$$

to get

$$\begin{aligned} &\sum_{\{i,k_1,\ldots,k_s\}} (x_i - y_\alpha)^{s-2} \prod_{j=i}^{s} \frac{1}{x_{k_j} - x_i} \\ &= \sum_{\{i,k_1,\ldots,k_s\}} \sum_{t=0}^{s-2} \binom{s-2}{t} x_i^t (-1)^{s-2-t} y_\alpha^{s-2-t} (-1)^s \\ &\qquad \times \frac{V^{(s)}(x_{k_1}, \ldots, x_{k_s})}{V^{(s+1)}(x_{k_1}, \ldots, x_{k_s}, x_i)} \\ &= \sum_{t=0}^{s-2} \binom{s-2}{t} (-1)^{s-2-t} y_\alpha^{s-2-t} (-1)^s \frac{W^{(s,t)}(x_{k_1}, \ldots, x_{k_s}, x_i)}{V^{(s+1)}(x_{k_1}, \ldots, x_{k_s}, x_i)} \end{aligned} \tag{15}$$

where

$$W^{(s,t)}(x_{k_1},\ldots,x_{k_s},x_i)=\det\begin{pmatrix}1&1&\ldots&1&1\\x_{k_1}&x_{k_2}&\ldots&x_{k_s}&x_i\\\vdots&\vdots&&\vdots&\vdots\\x_{k_1}^{s-1}&x_{k_2}^{s-1}&\ldots&x_{k_s}^{s-1}&x_i^{s-1}\\x_{k_1}^t&x_{k_2}^t&\ldots&x_{k_s}^t&x_i^t\end{pmatrix}. \quad (16)$$

We can now conclude, remarking that since $0\le t\le s-2$ two rows of the det W are the same and thus the whole expression is zero. We have thus proved that the wavefunction Eq. (4) is an eigenstate of the Hamiltonian Eq. (1) with the energy Eq. (12) for a mixture of fermions and bosons with SU($m|n$) spin symmetry.

We expect that this wavefunction is the ground state of the system in the supersymmetric case. First of all, without bosons, our wavefunction reduces to the ground state wavefunction for the multi-component fermionic system studied before [6]. Second, when both M and N are nonzero, it is impossible to construct lower energy eigenstates by using the lowering operators, in the same way as for the non-uniform supersymmetric t-J model [4, 7]. There are $M+N$ lowering operators, a_i, $i=1,2,\ldots,Q$ [5]. In fact, we have checked explicitly the following relations:

$$\left(\sum_{i=1}^{M}a_i\right)|\Psi\rangle=0,\qquad\left(\sum_{i=1}^{M}a_i\sigma_i^z\right)|\Psi\rangle=0,$$
$$\left(\sum_{i=M+1}^{Q}a_i\right)|\Psi\rangle=0,\qquad\left(\sum_{i=M+1}^{Q}a_i\tau_{i-M}^z\right)|\Psi\rangle=0, \quad (17)$$

where Ψ is the wavefunction Eq. (4). These two properties support our conclusion that the wavefunction we have constructed for the supersymmetric SU($m|n$) case is indeed the ground state in the subspace where the number of particles of each flavor is fixed.

Having studied the ground state wavefunction, we can now construct a set of non-trivial excited states. These states are obtained by multiplying the ground state wavefunction by a symmetrical product F_I of Hermite polynomials. Without bosons, the following excited states will reduce to those previously obtained [6]. Our wavefunctions take the form of

$$\Phi_I=\Psi\cdot F_I=\Psi\cdot\sum_{\substack{m_1,\ldots,m_Q\\m_1+\cdots+m_Q=I}}\prod_{i=1}^{Q}\frac{1}{m_i!}H_{m_i}(\sqrt{\omega}q_i),\qquad I=0,1,2,\ldots,\infty \quad (18)$$

where Ψ is the ground state given by Eq. (4) and the m_i are positive integers whose sum is the quantum number I. To prove that this wave functon is an eigenfunction of our Hamiltonian, we simply use the fact that Ψ is the ground state to get the following eigenequation

$$\begin{aligned} \frac{H\Phi_I}{\Phi_I} &= E_G + \omega I - \frac{1}{F_I}\sum_{\substack{i,j=1\\ i>j}}^{M} \frac{\lambda + \delta_{\sigma_i\sigma_j}}{q_i - q_j}\left(\frac{\partial F_I}{\partial q_i} - \frac{\partial F_I}{\partial q_j}\right) \\ &\quad -\frac{1}{F_I}\sum_{\substack{i,j=M+1\\ i>j}}^{Q} \frac{\lambda}{q_i - q_j}\left(\frac{\partial F_I}{\partial q_i} - \frac{\partial F_I}{\partial q_j}\right) \\ &\quad -\frac{1}{F_I}\sum_{i=1}^{M}\sum_{j=M+1}^{Q} \frac{\lambda}{q_i - q_j}\left(\frac{\partial F_I}{\partial q_i} - \frac{\partial F_I}{\partial q_j}\right). \end{aligned} \tag{19}$$

The three last terms are zero since

$$\frac{\partial F_I}{\partial q_i} = \frac{\partial F_I}{\partial q_j} \tag{20}$$

for all i and j. The corresponding energy spectrum is thus equally spaced and given by

$$E = E_G + \omega I. \tag{21}$$

Following standard arguments, it is expected that Eq. (21) gives the full energy spectrum. However due to possible degeneracies, the wavefunctions Eq. (18) do not yield the full set of excited states of the system.

In summary, we have considered the supersymmetric long range model confined in a harmonic potential. The ground state wavefunction and the ground state energy have been provided by us. We have also constructed a set of non-trivial excited states of Jastrow form. In the limit case of no bosons, our results reduce to those obtained previously for the SU(n) fermionic gas confined in harmonic potential. For the supersymmetric model studied here, the system still has lowering and raising operators, and the full energy spectrum of this system is also equal spaced. In fact, assuming that the degeneracy of each energy level is the same as if there were no interaction between the particles, one could construct the thermodynamics of the system very easily.

Acknowledgments

This work was supported in part by the Swiss National Science Foundation.

References

[1] F. Calogero, J. Math. Phys. **10**, 2191 (1969); **10**, 2197 (1969); **12**, 419 (1971).

[2] B. Sutherland, J. Math. Phys. **12**, 246, (1971); **12**, 251 (1971); Phys. Rev. A **4**, 2019 (1971); **5**, 1372 (1971).

[3] N. Kawakami and N. Kuramoto, Phys. Rev. B **50**, 4664 (1994).

[4] C. Gruber and D. F. Wang, Phys. Rev. B **50**, 3103 (1994).

[5] A. P. Polychronakos, Phys. Rev. Lett. **69**, 703 (1992); L. Brink, T. H. Hansson and M. A. Vasiliev, Phys. Lett. B **286**, 109 (1992).

[6] K. Vacek, A. Okiji and N. Kawakami, Phys. Rev. B **49**, 4637 (1992); J. Phys. A **7**, L201 (1994).

[7] J. T. Liu and D. F. Wang, Int. J. Mod. Phys. B **10**, 3685 (1996).

[8] T. Yamamoto and O. Tsuchiya, J. Phys. A **29**, 3977 (1996).

$1/r^2$ *t-J* MODEL IN A MAGNETIC FIELD

James T. Liu
Department of Physics
The Rockefeller University
1230 York Avenue
New York, NY 10021-6399

D. F. Wang
Institut de Physique Théorique
Ecole Polytechnique Fédérale de Lausanne
PHB-Ecublens, CH-1015 Lausanne
Switzerland

Abstract We study the one-dimensional supersymmetric t-J model with $1/r^2$ interaction threaded by magnetic flux. Because of the long-range interaction, the effect of this flux leads to a modification of the electron hopping term. We present an exact solution of this model for all values of the flux, concisely formulated as a set of Bethe-ansatz-like equations. Examination of the ground state shows that the persistent currents at zero temperature do not exhibit a parity effect despite the fact that the long-range t-J model falls in the Luttinger-liquid universality class. This exception to Leggett's conjecture arises because of the special nature of the long-range hopping.

Exact solutions have provided us with an interesting way to deal nonperturbatively with systems of strongly correlated electrons. Notable examples are the electron systems with δ-function interaction [1], the Hubbard model [2], and the short-range t-J model [3]. These models are solvable by Bethe ansatz and have played an important role in understanding the physics of the one-dimensional electron gas. A particularly interesting class of lattice models that are exactly solvable despite long-range interactions are the Haldane-Shastry spin chain of $1/r^2$ exchange interaction [4, 5] and its many variations. The latter include the supersymmetric t-J models of long range hopping and exchange [6, 7, 8, 9, 10, 11] as well as multicomponent generalizations. Since the closed spin chain admits a Bethe-ansatz-like solution, this indicates that

Reprinted from Liu *et al.*, Phys. Rev. B 56 (1997) 2312-2315 the quasiparticle interactions are statistical in nature and arise from demanding periodicity of the chain.

While it is natural to close the chain by imposing periodic boundary conditions, we may also consider the case where the closed chain is threaded by magnetic flux. Due to the nontrivial topology, electrons transported around the chain then pick up an Aharonov-Bohm phase in the presence of a nonzero flux. In models with nearest-neighbor exchange it is straightforward to encode this phase (and hence the flux) by imposing twisted boundary conditions on the wave functions. However such a prescription needs to be modified in the presence of long-range interactions where particles may hop between any arbitrary pair of sites. We show how this may be done for the supersymmetric t-J model with inverse square interaction and investigate its ground state and full energy spectrum. Based on previous results in the absence of magnetic flux [10], we derive a set of Bethe-ansatz-like equations appropriate to twisted boundary conditions.

We consider a system of electrons on a one-dimensional ring described by the SU(2) supersymmetric t-J model. For a uniform and closed chain of L lattice sites, the Hamiltonian takes the form

$$\begin{aligned} H_{tJ} &= -\frac{1}{2}\sum_{\sigma=\uparrow,\downarrow}\sum_{1\le l\ne m\le L}[t(l-m)c^{\dagger}_{l\sigma}c_{m\sigma}+\text{H.c.}] \\ &+\frac{1}{2}\sum_{1\le l\ne m\le L}J(l-m)[P^{\sigma}_{lm}-(1-n_l)(1-n_m)]\,, \end{aligned} \tag{1}$$

where the hopping strength, $t(n)$, and exchange interaction, $J(n)$, are functions only of separation due to the rotational invariance of the ring. P^{σ}_{lm} is the spin exchange operator and n_l is the electron number operator. We have implicitly assumed a projection onto single occupancy at each site.

Without magnetic flux, the supersymmetric long-range model has an interaction strength given by $t(n) = J(n) = 1/d^2(n)$ where $d(n) = (L/\pi)\sin(\pi n/L)$. While $d(n)$ is most readily interpreted as the chord length between sites n units apart, this interpretation is not easily generalized to encompass nonzero flux. In particular, an electron hopping along a chord must travel through the interior of the ring where it would be sensitive to the actual magnetic field and not just the flux. To avoid this difficulty, we use the alternate interpretation of $J(n)$ as the periodic version of $1/n^2$, namely [12, 11, 13]

$$J(n) = \sum_{k=-\infty}^{\infty}\frac{1}{(n+kL)^2} = \left(\frac{\pi}{L}\right)^2\frac{1}{\sin^2(\pi n/L)}\,, \tag{2}$$

which represents the sum of hopping over all multiples of the period L with periodic boundary conditions. It is now straightforward to generalize this to twisted boundary conditions appropriate to a ring threaded by flux. We introduce a dimensionless flux ϕ, represented by the vector potential $A = \phi(\phi_0/L)$ where $\phi_0 = hc/e$, so that electrons pick up a phase $e^{2\pi i\phi}$ when transported once around the ring. In this case, the hopping interaction is twisted and becomes

$$t_\phi(n) = \sum_{k=-\infty}^{\infty} \frac{e^{2\pi i\phi(n+kL)/L}}{(n+kL)^2} . \tag{3}$$

Since the exchange interaction is insensitive to the flux, $J(n)$ is still given by Eq. (2). This interaction was introduced by Fukui and Kawakami for the twisted Haldane-Shastry model in Ref. [13] where the sum was carried out for rational twists, $\phi = p/q$. It turns out, however, that this infinite sum can actually be evaluated for arbitrary ϕ, yielding the result

$$t_\phi(n) = \left(\frac{\pi}{L}\right)^2 \frac{z^{n\lfloor\phi\rfloor}\{[1-(\phi-\lfloor\phi\rfloor)] + (\phi-\lfloor\phi\rfloor)z^n\}}{\sin^2(\pi n/L)} , \tag{4}$$

where $z = e^{2\pi i/L}$ and $\lfloor\phi\rfloor$ is the greatest integer not exceeding ϕ. This expression is remarkable for the fact that it is piecewise *linear* and continuous in ϕ, even though the flux originally entered in the exponent of Eq. (3). We note that $t_\phi(n)$ satisfies the conditions $t_{-\phi}(n) = t_\phi(n)^*$ and $t_{-\phi}(-n) = t_\phi(n)$, which is readily apparent from Eq. (3), but hidden in the summed expression of Eq. (4). To simplify our subsequent discussion, we may restrict ϕ to lie in the range $0 \le \phi \le 1$, obviating the need for the greatest integer function. Other values of the flux may always be brought into this range by a gauge transformation with resultant shift in the lattice momentum. In this case, Eq. (4) becomes

$$t_\phi(n) = J(n)[(1-\phi) + \phi z^n] , \tag{5}$$

demonstrating that the effect of an arbitrary flux is to simply give a linear interpolation between different systems, each with an integral value of ϕ.

Previous techniques for solving the Haldane-Shastry and t-J models without flux [4, 6, 10] are easily extended to the present case, given by the twisted hopping $t_\phi(n)$. In particular, the t-J model may be solved by introducing a basis of Jastrow wave functions describing the down-spin and hole excitations about a fully polarized up-spin background. While this background may appear unnatural in the presence of flux, it nevertheless allows an immediate generalization of the exact solution constructed in Ref. [10]. In order to apply the techniques of Ref. [10], we need the sum formula

$$\begin{aligned}\sum_{n=1}^{L} z^{Jn}(1-z^n)^a t_\phi(n) \\ = 2\frac{\pi^2}{L^2}\{\delta_{a,0}[\tfrac{1}{6}(L^2-1) + \phi(1-\phi) - (J+\phi)(L-(J+\phi))] \\ +\delta_{a,1}[L-2(J+\phi)-1] + 2\delta_{a,2}\} ,\end{aligned} \tag{6}$$

which follows from Eq. (5) and the zero-flux sum formula [4]. This expression holds for non-negative J whenever $0 \leq \phi \leq 1$ and $0 \leq a \leq L - 1 - J$ and generalizes the result for rational twists presented in Ref. [13] while reducing to the standard expression [4] when $\phi = 0$. These results indicate that the restriction to rational twists is unnecessary, so that there is no distinction between rational and irrational twist angles in this strictly one-dimensional system. We note that the ϕ^2 terms cancel in the above sum formula as they must; it is the vanishing of this quadratic term (which persists in the ground-state energy) that is ultimately responsible for the disappearance of the ground-state parity effect in this model.

Based on the fact that the three terms in Eq. (6) correspond to constant, two-body, and three-body terms in the Hamiltonian [4, 6], we see that the flux has no effect on the cancellation of three-body terms. This immediately shows that the quasiparticles remain free, up to statistical interactions, even in the presence of flux. From the constant and two-body terms, it is apparent that ϕ acts to shift the quasiparticle momenta, leading to a modified dispersion relation. Since this is the extent of the modification to the solution caused by nonzero ϕ, the results of Ref. [10] are easily extended to the case of twisted hopping.

For a spin chain with $M_\downarrow$ down spins and Q holes, we introduce two sets of pseudomomenta: p_i $(i = 1, 2, \ldots, M_\downarrow + Q)$ and q_α $(\alpha = 1, 2, \ldots, Q)$. The solution to this supersymmetric t-J model may then be written in a Bethe-ansatz-like form

$$\begin{aligned} p_i L &= 2\pi J_i + \sum_{j \neq i} \theta(p_i - p_j) - \sum_{\alpha} \theta(p_i - q_\alpha) , \\ \sum_i \theta(q_\alpha - p_i) &= 2\pi I_\alpha , \end{aligned} \tag{7}$$

where the step function $\theta(x) = \pi \,\mathrm{sgn}(x)$ is the (statistical) scattering phase. The fermionic quantum numbers J_i and I_α are either integers or half-integers and are restricted to lie in the ranges $|J_i| \leq (L - M_\downarrow + 1)/2$ and $-(M_\downarrow + Q)/2 \leq I_\alpha \leq (M_\downarrow + Q)/2 - 1$ respectively. Since the q_α's label the hole degrees of freedom, it gives rise to a natural splitting of the pseudomomenta $\{p_i\}$ into $M_\downarrow$ spin and Q hole degrees of freedom. Namely we take Q to be the set of pseudomomenta p_i satisfying

$$\sum_\alpha [\theta(p_i - q_\alpha) - \theta(p_{i-1} - q_\alpha)] = 2\pi \, . \tag{8}$$

There are exactly Q such pseudomomenta, corresponding to the hole excitations. The remaining $M_\downarrow$ pseudomomenta then correspond to spin excitations. Using this distinction, the energy spectrum and momentum of the system are given by

$$\begin{aligned} E(\phi) &= \frac{\pi^2}{6}L(1-1/L^2) + \sum_{i\notin Q} \epsilon_0(p_i) + \sum_{i\in Q} \epsilon_\phi(p_i) \,, \\ P(\phi) &= (L-1)\pi + 2\pi\phi(1-Q/L) - \sum_{i=1}^{M_\downarrow+Q} (p_i - \pi) \bmod 2\pi, \end{aligned} \tag{9}$$

where the single particle dispersion relation is

$$\begin{aligned} \epsilon_\phi(k) &= \tfrac{1}{2}[(k+2\pi\phi/L)^2 - \pi^2 + 4\pi^2\phi(1-\phi)/L^2] \\ &= (1-\phi)\epsilon_0(k) + \phi\epsilon_0(k+2\pi/L) \,, \end{aligned} \tag{10}$$

for $0 \le \phi \le 1$. This piecewise linear form of the dispersion relation follows directly from the behavior of the hopping term, Eq. (5). We wish to stress that, while this solution has the form of a Bethe ansatz, it was actually derived as an exact solution based on the construction of a complete basis of Jastrow wave functions and a proper ordering of the Hilbert space, as described in Refs. [12] and [10].

States in the excitation spectrum are labeled by individually nonoverlapping quantum numbers J_i and I_α. The J_i may be represented as a string of 0's and 1's of length $L - M_\downarrow$ with $M_\downarrow + Q$ occupied positions represented by 1's [14]. The I_i then in turn label which of these 1's correspond to hole excitations (and hence are sensitive to the flux). The interpretation of the Bethe-ansatz-like equations (7) is to separate the spin excitations by inserting a 0 before every spin excitation. The resulting string then specifies the pseudomomenta p_i, lying in the range $[-\pi, \pi)$. From Eq. (10), it is evident that states in the middle of the string have lowest energy. Therefore the ground state of the t-J model, in a sector of fixed $M_\downarrow$ and Q, has pseudomomenta of the general form $p_i \in (\ldots 0010101\underline{1111}0101000\ldots)$, with the hole excitations centrally located (and underlined). In order to study the ground-state properties, we introduce uniform spin and hole momenta, J_s and J_h (integral or half-integral as appropriate), perturbing the string of 1's to the left or right. In this case, the corresponding eigenenergies are

$$\begin{aligned} \frac{L^2}{\pi^2}E(J_s, J_h) &= E_0 + 2M_\downarrow J_s^2 + 2Q[(J_h+\phi)^2 \\ &\quad + (J_s + J_h + \phi)^2 - 2\phi^2] \,, \end{aligned} \tag{11}$$

where

$$\begin{aligned} E_0 &= \tfrac{1}{6}L(L^2-1) + \tfrac{2}{3}(Q+M_\downarrow)[(Q+M_\downarrow)^2 - 1] + 4\phi Q \\ &\quad - \tfrac{1}{2}(Q+M_\downarrow)L^2 - \tfrac{1}{2}Q(Q+M_\downarrow)(Q+2M_\downarrow) \,. \end{aligned} \tag{12}$$

Table 1. Exact ground-state energies $E(\phi)$ and momenta $P(\phi)$ of the t-J model with twisted boundary conditions for $0 \leq \phi \leq \frac{1}{2}$. Taking into account a level crossing at $\phi = \frac{1}{2}$, the absolute ground state for $\frac{1}{2} \leq \phi \leq 1$ instead has energy $E(1-\phi)$ and momentum $-P(1-\phi)$.

N_e mod 4	$E(\phi)$	$P(\phi)$
0	$E_g + \pi^2 n_h/L$	$P_g + \pi(1+n_h)$
1	$E_g + \pi^2[2 + n_h(1+8\phi)]/4L$	$P_g - \frac{\pi}{2}(1 - n_h - 1/L)$
2	$E_g + 4\pi^2 \phi n_h/L$	P_g
3	$E_g + \pi^2[2 + n_h(1+8\phi)]/4L$	$P_g - \frac{\pi}{2}(1 - n_h + 1/L)$

For fixed $M_\downarrow$ and Q, the ground state has both J_s and $J_h + \phi$ as close to zero as possible. These states correspond to exact Jastrow product wave functions describing the ground state as well as uniform excitations of the t-J model.

To further examine the ground state of this model, we work at a fixed hole fraction, $n_h \equiv Q/L$. Denoting the number of electrons by $N_e = M_\uparrow + M_\downarrow = L - Q$, the ground state is either an $SU(2)$ singlet for even N_e or a doublet for odd N_e. Due to finite-size effects, the ground-state properties depend on the value of N_e mod 4. At zero flux, whenever $N_e \neq 2$ mod 4, the ground state carries momentum and is hence two-fold degenerate. However this degeneracy is always lifted for nonzero ϕ which breaks time reversal symmetry. The exact ground-state energies and momenta are given in Table 1, where the bulk quantities are

$$\begin{aligned} E_g &= -\frac{\pi^2 L}{12}[n_h(3 - n_h^2) + 2(3 + 2n_h)/L^2]\,, \\ P_g &= 2\pi\phi(1 - n_h)\,. \end{aligned} \tag{13}$$

We note that the linear spectral flow apparent from the cancellation of ϕ^2 terms in Eq. (10) indicates that the ground state at zero flux is always the absolute lowest energy state since any flow to lower energy has nowhere to terminate. Hence the ground state is always diamagnetic, regardless of the number of electrons.

It is well known that the ground state of a noninteracting one-dimensional electron gas is either diamagnetic or paramagnetic, depending on whether the number of electrons in the system is even or odd. Leggett has conjectured that this parity effect persists in the presence of interactions [15]; this has subsequently been proven by Loss for generic Luttinger liquids [16]. Thus it is somewhat of a surprise that this long range t-J model provides an exception to Leggett's conjecture despite the fact that its low-lying physics is still described by a Luttinger-liquid fixed-point Hamiltonian. The origin of this breakdown may be traced to the unusual quasiparticle dispersion relation, Eq. (10), which in turn is an artifact of the long-range nature of the electron hopping interaction.

Therefore this model does not invalidate Leggett's conjecture, but rather emphasizes the peculiar features of long-range hopping in the presence of a magnetic field, as was already apparent in the subtleties in constructing $t_\phi(n)$ and its resulting linearity in ϕ. This shows that there are sharp differences between models with and without long range hopping in the presence of a magnetic field.

We now turn to a generalization to the SU($1|K$) supersymmetric t-J model with long-range interactions. Since the two-body nature of the quasiparticle interactions is unaffected by the flux, we may approach the SU($1|K$) generalization via the asymptotic Bethe ansatz (ABA), which was constructed at zero flux in Refs. [7] and [17]. Since the magnetic flux twists all K fermionic species identically, it is natural to write the ABA in terms of fermionic excitations above the purely bosonic vacuum (which we denote $F^K B$). For this choice of grading, only the first nesting is affected by the flux. To proceed, we introduce K sets of pseudomomenta

$$p_i^{(a)}: \quad i = 1, 2, \ldots, N_a , \tag{14}$$

where $a = 1, 2, \ldots, K$ and $N_a = \sum_{i=a}^{K} M_i$ (M_i is the number of electrons with spin component i). Note that N_1 gives the total number of SU(K) electrons. These quasimomenta satisfy the following ABA equations:

$$\begin{aligned}
p_i^{(1)} L &= 2\pi J_i + \sum_\alpha \theta(p_i^{(1)} - p_\alpha^{(2)}) , \\
\sum_i \theta(p_\alpha^{(2)} - p_i^{(1)}) &= 2\pi I_\alpha^{(2)} + \sum_\beta \theta(p_\alpha^{(2)} - p_\beta^{(2)}) - \sum_\gamma \theta(p_\alpha^{(2)} - p_\gamma^{(3)}) , \\
&\ \ \vdots \\
\sum_\gamma \theta(p_\alpha^{(K)} - p_\gamma^{(K-1)}) &= 2\pi I_\alpha^{(K)} + \sum_\beta \theta(p_\alpha^{(K)} - P_\beta^{(K)}) .
\end{aligned} \tag{15}$$

The quantum numbers $\{J_i\}, \{I_\alpha^{(2)}\}, \ldots, \{I_\alpha^{(K)}\}$ are integers or half-integers which are distinct within each set, respectively. This set of equations is unchanged from the case without flux [7, 17]; the only place where ϕ enters is in the energy and momentum, given by

$$\begin{aligned}
E(\phi) &= -\frac{\pi^2}{6} L(1 - 1/L^2) - \sum_{i=1}^{N_1} \epsilon_\phi(p_i^{(1)}) , \\
P(\phi) &= \sum_{i=1}^{N_1} (p_i^{(1)} + 2\pi\phi/L - \pi) \bmod 2\pi .
\end{aligned} \tag{16}$$

This provides the exact energy spectrum of the SU(K) t-J model in the presence of flux ϕ, even in the nonasymptotic regime.

For the ordinary SU(2) t-J model threaded by flux, the ABA equations given in the F^2B grading provide a more symmetrical description than the microscopically derived equations, (9). Nevertheless, the BF^2 picture of Eq. (7) has an advantage in that the complete level degeneracies are understood [10] independent of the SU(1|2) supermultiplet structure, which is spontaneously broken for non-zero flux.

In summary, we have provided an exact solution to the long-range t-J model in the presence of arbitrary flux. This solution indicates that there is no meaningful difference between rational and irrational values of the flux. We have also given solutions for the full energy spectra of the general SU(K) t-J model based on an asymptotic Bethe ansatz. It should be feasible to prove these solutions exact by studying the system with a complete basis of Jastrow wave functions as was done for the SU(2) model. For this latter case, the form of the exact solutions, only slightly changed in the presence of flux, indicates that this model remains integrable, even though the manifest SU(1|2) supersymmetry has been lost. Thus it is apparent that part of the Yangian symmetry [18] remains. Work on finding an infinite set of commuting constants of motion in the presence of magnetic flux is currently in progress.

Note added: Recently, we bacame aware of Ref. [19], which independently addressed many of the same issues. B. Sutherland also recently informed us that the linear dependence of the ground-state energy on the flux can be derived for the twisted t-J model by imposing twisted boundary conditions on the continuum model [20, 21], and then by taking the strong interaction limit to decouple the lattice oscillation modes from the internal degrees of freedom at each site. This approach should enable one to handle the SU($1|K$) t-J model with twisted boundary conditions in a straightforward manner.

Acknowledgments

We wish to thank C. Gruber, H. Kunz, R. Khuri and X. Q. Wang for stimulating discussions. In addition, we wish to thank C. A. Stafford for fruitful discussions on mesoscopic systems and H. C. Ren for suggesting how to sum Eq. (3). This work was supported in part by the U.S. Department of Energy under Grant No. DOE-91ER40651-TASKB, and by the Swiss National Science Foundation.

References

[1] C. N. Yang, Phys. Rev. Lett. **19**, 1312 (1967).

[2] E. H. Lieb and F. Y. Wu, Phys. Rev. Lett. **20**, 1445 (1968); C. N. Yang, *ibid.* **63**, 2144 (1989); C. N. Yang and S. C. Zhang, Mod. Phys. Lett. B **4**, 759 (1990).

[3] B. Sutherland, Phys. Rev. B **12**, 3795 (1975).

[4] F. D. M. Haldane, Phys. Rev. Lett. **60**, 635 (1988).

[5] B. S. Shastry, Phys. Rev. Lett. **60**, 639 (1988).

[6] Y. Kuramoto and H. Yokoyama, Phys. Rev. Lett. **67**, 1338 (1991).

[7] N. Kawakami, Phys. Rev. B **46**, 1005 (1992).

[8] D. F. Wang and C. Gruber, Phys. Rev. B **49**, 15 712 (1994).

[9] C. Gruber and D. F. Wang, Phys. Rev. B **50**, 3103 (1994).

[10] D. F. Wang, J. T. Liu and P. Coleman, Phys. Rev. B **46**, 6639 (1992).

[11] Z. N. C. Ha and F. D. M. Haldane, Phys. Rev. B **46**, 9359 (1992).

[12] B. Sutherland J. Math. Phys. (N.Y.) **12**, 246 (1971); **12**, 251 (1971); Phys. Rev. A **4**, 2019 (1971); **5**, 1372 (1972).

[13] T. Fukui and N. Kawakami, Phys. Rev. Lett. **76**, 4242 (1996).

[14] F. D. M. Haldane, Phys. Rev. Lett. **66**, 1529 (1991).

[15] A. J. Leggett, in *Granular Nanoelectronics*, edited by D. K. Ferry *et al.*, NATO ASI Series B, Vol 251 (Plenum, New York, 1991), p. 297.

[16] D. Loss, Phys. Rev. Lett. **69**, 343 (1992).

[17] J. T. Liu and D. F. Wang, Int. J. Mod. Phys. B **10**, 3685 (1996); D. F. Wang and J. T. Liu, Phys. Rev. B **54**, 584 (1996).

[18] F. D. M. Haldane, in *Proceedings of the 16th Taniguchi Symposium on Condensed Matter Physics, Kashikojima, Japan*, edited by A. Okiji and N. Kawakami (Springer, Berlin, 1994).

[19] T. Fukui and N. Kawakami, Phys. Rev. B **54**, 5346 (1996).

[20] B. Sutherland and B. S. Shastry, Phys. Rev. Lett. **71**, 5 (1993).

[21] B. Sutherland, R. A. Römer and B. S. Shastry, Phys. Rev. Lett. **73**, 2154 (1994).

INTERACTION-INDUCED ENHANCEMENT AND OSCILLATIONS OF THE PERSISTENT CURRENT

C. A. Stafford
Institut de Physique Théorique
Université de Fribourg
CH-1700 Fribourg
Switzerland

D. F. Wang
Institut de Physique Théorique
EPF Lausanne
CH-1015 Lausanne
Switzerland

Abstract The persistent current I in integrable models of multichannel rings with both short- and long-ranged interactions is investigated. I is found to oscillate in sign and increase in magnitude with increasing interaction strength due to interaction-induced correlations in the currents contributed by different channels. For sufficiently strong interactions, the contributions of all channels are found to add constructively, leading to a giant enhancement of I. Numerical results confirm that this parity-locking effect is robust with respect to intersubband scattering.

The absence of macroscopic persistent currents in normal metal rings even at $T = 0$ is a consequence of a parity effect stemming from Fermi statistics, first discussed by Byers and Yang [1]: For N spinless electrons in a purely one-dimensional (1D) ring, the persistent current I is diamagnetic if N is even and paramagnetic if N is odd, independent of disorder and interactions [2, 3], and takes a maximum value $I_0 = ev_F/L$ for a clean ring, where v_F is the Fermi velocity and L the circumference of the ring. In a ring with many independent channels, I is the sum of many such diamagnetic and paramagnetic contributions, and is thus very small [1]. Büttiker, Imry, and Landauer [4] argued that persistent currents should, nonetheless, exist at the mesoscopic level, and calculations with noninteracting electrons [5] predicted typical currents of

Reprinted from Stafford *et al.*, Phys. Rev. B 56 (1997) R4383-R4386 order $I_0\ell/L$ for a diffusive metallic ring, where $\ell \ll L$ is the elastic mean free path. Subsequently, persistent currents were observed experimentally in both normal metal [6, 7] and semiconductor [8, 9] rings. Surprisingly, the persistent currents observed in metallic rings were roughly two orders of magnitude larger than those predicted by theories neglecting electron-electron interactions [5, 10], being of order I_0 in individual rings [7]. Mean-field calculations [11, 12, 13] for multichannel rings, renormalization group results for 1D rings [14], and exact diagonalization studies [15] found that the persistent current in disordered systems can be enhanced by repulsive interactions due to the suppression of charge fluctuations, but the very large experimentally observed magnitude of I is still generally considered to be a mystery.

In this paper, we propose a nonperturbative interaction effect which can lead to a large enhancement of I in multichannel rings, *even in the ballistic regime*, due to the fact that the parities of different channels are no longer independent in an interacting system. We investigate two integrable models of interacting M-channel rings, $SU(M)$ fermions with inverse-square $V(x) = g/d(x)^2$ and delta-function $V(x) = U\delta(x)$ interactions. In the prior model, we show that for $g > 0$ the persistent currents of all channels add constructively provided $k_F L > 2\pi M$, leading to a large persistent current whose parity depends only on the total number of electrons N. In the $SU(M)$ delta-function gas, I is found to oscillate in sign and increase in magnitude with increasing U due to a progressive condensation of electrons into the lowest subband, leading to a parity-locking effect for $U > 16M^4\hbar v_F/3\pi$, where v_F is the Fermi velocity. A disordered two-channel ring with interchannel interactions is also investigated numerically, and shows, importantly, that the parity-locking effect persists even when the subbands are mixed strongly by disorder. Qualitatively, the parity-locking effect arises in a thin ring because for sufficiently strong repulsive interactions, electrons can no longer pass each other. The elimination of transverse nodes in the many-body wave function (which are necessary for two electrons to pass each other) leads to a state whose parity depends only on the total number of electrons, as discussed by Leggett [2], and whose persistent current is consequently large.

Spinless electrons in a nondisordered ring of arbitrary cross section with M transverse channels, threaded by an Aharanov-Bohm flux $(\hbar c/e)\phi$, may be represented by 1D $SU(M)$ fermions. The transverse degrees of freedom may be represented by an $SU(M)$ spin variable $\sigma = 1, \ldots, M$. Disorder and interactions will in general lead to intersubband scattering, which breaks this $SU(M)$ symmetry. In order to preserve the integrability of the model, we consider a special class of interactions without intersubband scattering, i.e., interactions which depend only on the electrons' coordinates along the ring. Intersubband scattering will be shown not to alter the results obtained for these integrable models in a fundamental way. The Hamiltonian of the system is

$$H = -\frac{1}{2}\sum_{i=1}^{N}\frac{\partial^2}{\partial x_i^2} + \sum_{i<j} V(x_i - x_j) + \sum_{\sigma=1}^{M} K_\sigma \varepsilon_\sigma, \tag{1}$$

where K_σ is the number of electrons in subband σ, $N = \sum_\sigma K_\sigma$, and ε_σ is the energy minimum of subband σ. Units with $\hbar = m = 1$ are used. The Aharanov-Bohm flux leads to the twisted boundary condition [1]

$$\Psi(x_1\sigma_1, \ldots, (x_i + L)\sigma_i, \ldots, x_N\sigma_N) = e^{i\phi}\Psi(x_1\sigma_1, \ldots, x_i\sigma_i, \ldots, x_N\sigma_N). \tag{2}$$

For simplicity, let us consider equally spaced subbands $\varepsilon_{\sigma+1} - \varepsilon_\sigma = \Delta \equiv E_F/M$. The subband splitting Δ plays the role of an $SU(M)$ magnetic field. As we shall see, the effect of repulsive interactions is to renormalize this effective field, causing a condensation of electrons into the lowest subband.

At $T = 0$, the equilibrium persistent current is given by $I(\phi) = -(e/\hbar)\partial E_0/\partial\phi$, where $E_0(\phi)$ is the ground state energy of Eq. (1), subject to the boundary condition (2). I is a periodic function of ϕ with period 2π, and may thus be expressed as a Fourier series,

$$I(\phi) = \sum_{n=1}^{\infty} I_n \sin(n\phi). \tag{3}$$

The value of I at $\phi = \pi/2$ (1/4 flux quantum) is determined by the odd harmonics, $I(\phi_0/4) = I_1 - I_3 + I_5 - \cdots$, and may be taken as a measure of the first harmonic, assuming the higher odd harmonics are small. For $V(x) = 0$, one finds

$$I(\phi_0/4) = \frac{eh\pi}{2mL^2}\sum_{\sigma=1}^{M}(-1)^{K_\sigma}K_\sigma. \tag{4}$$

For $M \gg 1$, this leads to the well-known [5] result $|I| \sim M^{1/2}I_0$ due to the random parities of the different channels. The system may be either diamagnetic or paramagnetic, depending on the channel occupancies K_σ.

Let us next consider a model with long-range interactions: $V(x) = g/d(x)^2$, where $d(x) = (L/\pi)|\sin(\pi x/L)|$ is the chord length along the ring. This model was introduced and solved by Sutherland [16] for the case $M = 1$ and $\phi = 0$, and can be shown to be integrable [17] for arbitrary M, ϕ, and $g \geq 0$. For $g > 0$, the ground state is highly degenerate in the limit $L \to \infty$ in the absence of $SU(M)$ symmetry breaking ($\Delta = 0$) due to the strong repulsion of the potential at the origin, which prohibits particle exchange. In a finite ring, these states differ in energy by at most $\pi\hbar v_F/L$ due to boundary effects; all electrons will thus be condensed into the lowest subband for $\Delta > \pi\hbar v_F/L$, i.e., for $k_F L > 2\pi M$, which is satisfied provided the ring is sufficiently thin. The ground state of the system in this "ferromagnetic" state has the Jastrow product form

$$\Psi(\{x\}) = \exp\left(i\frac{\phi - a}{L}\sum_{k=1}^{N} x_k\right) \prod_{1\le i<j\le N} \left|\sin\left(\frac{x_i - x_j}{L}\pi\right)\right|^{\lambda} \sin\left(\frac{x_i - x_j}{L}\pi\right) \tag{5}$$

for $0 \le \phi \le \pi$, where $a = 0$ if N is odd and $a = \pi$ if N is even. Here $\lambda = \sqrt{g + 1/4} - 1/2$. One readily verifies that Ψ is an eigenstate of Eq. (1), has the correct symmetry, and obeys the twisted boundary condition (2). This eigenstate is a positive vector when the particles are ordered, and is therefore the ground state of the system. The ground state energy is found to be

$$E_0(\phi) = \frac{\pi^2(\lambda+1)^2 N(N^2-1)}{6L^2} + \frac{N}{2}\left(\frac{\phi - a}{L}\right)^2. \tag{6}$$

The corresponding persistent current is

$$I(\phi_0/4) = (-1)^N \frac{e\hbar\pi N}{2mL^2} \sim (-1)^N M I_0. \tag{7}$$

The condensation of all electrons into the lowest subband caused by the strong repulsive interactions thus leads to an enhancement of the typical persistent current by a factor of $M^{1/2}$ in the ballistic regime, due to the suppression of the parity effect of Byers and Yang.

Let us next consider a model with short-ranged interparticle interactions $V(x) = U\delta(x)$. This system is integrable for arbitrary U, and the eigenenergies with twisted boundary conditions may be determined from a straightforward generalization of the nested Bethe ansatz of Sutherland [18] to the case $\phi \neq 0$. The energy of the system may be expressed as

$$E = \sum_{j=1}^{N} k_j^2/2 + \sum_{\sigma=1}^{M} K_\sigma \varepsilon_\sigma, \tag{8}$$

where the pseudomomenta k_j are a set of N distinct numbers which satisfy the coupled equations

$$\exp[i(Lk_j - \phi)] = \prod_{\alpha=1}^{N-K_1} \frac{k_j - \Lambda_\alpha^{(1)} + iU/2}{k_j - \Lambda_\alpha^{(1)} - iU/2}, \tag{9}$$

$$\prod_{\beta=1}^{N_{n-1}} \frac{\Lambda_\alpha^{(n)} - \Lambda_\beta^{(n-1)} - iU/2}{\Lambda_\alpha^{(n)} - \Lambda_\beta^{(n-1)} + iU/2} \prod_{\gamma=1}^{N_{n+1}} \frac{\Lambda_\alpha^{(n)} - \Lambda_\gamma^{(n+1)} - iU/2}{\Lambda_\alpha^{(n)} - \Lambda_\gamma^{(n+1)} + iU/2}$$
$$= -\prod_{\beta=1}^{N_n} \frac{\Lambda_\alpha^{(n)} - \Lambda_\beta^{(n)} - iU}{\Lambda_\alpha^{(n)} - \Lambda_\beta^{(n)} + iU}, \qquad n = 1, \ldots, M-1, \tag{10}$$

where $\Lambda_\alpha^{(n)}$, $\alpha = 1, \ldots, N_n = N - \sum_{\sigma=1}^{n} K_\sigma$ are distinct numbers, with $\Lambda_j^{(0)} = k_j$. For $\Delta = 0$, the ground state is an $SU(M)$ singlet when N is an odd multiple of M. As Δ increases, electrons in the higher subbands are transferred to lower subbands, until all electrons are condensed into the lowest subband for $\Delta > \Delta_c$. This phenomenon is analogous to the spin-polarization transition of the 1D Hubbard model in a magnetic field, studied by Carmelo *et al.* and by Frahm and Korepin [19]. Using the techniques of Ref. [19], one finds

$$\Delta_c = (1/4\pi)[U^2 + (2\pi n)^2] \tan^{-1}(2\pi n/U) - Un/2, \tag{11}$$

where $n = N/L$. For $\Delta > \Delta_c$, the system is in a parity-locked state, with persistent current given by Eq. (7). It is useful to consider some limiting cases of Eq. (11): Using $k_F \simeq \pi n/M$, one finds $\Delta_c \simeq M^2 E_F$ for $U = 0$ and $\Delta_c \simeq 16 M^3 E_F v_F / 3\pi U$ for $U/v_F \gg M$. For fixed Δ, the critical interaction strength required to enforce parity-locking is thus

$$U_c \simeq 16 M^4 v_F / 3\pi. \tag{12}$$

For $M \gg 1$, very strong interactions are thus required to cause complete parity-locking, which would lead to a *macroscopic* persistent current. This result is in contrast to that for the preceding model, which exhibited parity-locking for any value of the interaction $g > 0$, provided the ring was sufficiently thin.

In order to see how the parity-locking effect develops as a function of interaction strength, let us consider the simple case of a two-channel ring with N odd; then the parities of the two channels are necessarilly opposite. For mesoscopic rings, $SU(2)$ excitations of the type considered by Kusmartsev [20] and by Yu and Fowler [21], which lead to a ϕ_0/N periodicity of the persistent current, can be neglected [21]. The persistent current for $M = 2$ is then given by

$$I(\phi_0/4) = (-1)^{K_1}(K_1 - K_2) I_0/N. \tag{13}$$

For $\Delta \ll \Delta_c$, the polarization $(K_1 - K_2)/L \simeq 2\chi\Delta$, where the susceptibility χ may be evaluated from Eqs. (8)–(10) using the method of Shiba [22]; one obtains $\chi = 2/\pi v_F$ for $U = 0$ and $\chi \simeq 3U/(2\pi v_F)^2$ for $U \gg v_F$. The magnitude of the persistent current is thus given by $|I| \simeq 4e\Delta/\pi\hbar N$ for $U = 0$, and is increased by the factor $3U/8\pi v_F$ for $v_F \ll U \ll U_c$. As U is increased, the persistent current thus oscillates in sign and grows roughly linearly in magnitude due to the progressive transfer of electrons from the upper to the lower subband. For $M \gg 1$, the evolution of the system toward the parity-locked state as U is increased will of course be more complicated, but one nonetheless expects I to fluctuate in sign and increase in magnitude as electrons condense into the lowest subband.

A peculiarity of the integrable models considered to this point is that the number of electrons in each channel is a constant of the motion. Both disorder and more realistic interactions which depend on the transverse coordinate will break this symmetry, and it is therefore important to verify that the parity-locking effect is not destroyed. To this end, we have considered a disordered two-channel ring, modeled in the tight-binding approximation, with a nearest-neighbor interchain interaction V included to induce interchannel correlations [23]. The Hamiltonian is

$$\begin{aligned} H &= \sum_{i=1}^{L}\Bigg[\sum_{\alpha=1}^{2}\left(e^{i\phi/L}c_{i\alpha}^{\dagger}c_{i+1\alpha}+\text{H.c.}+\varepsilon_{i\alpha}\rho_{i\alpha}\right) \\ &\quad +\frac{\Delta}{2}\left(c_{i1}^{\dagger}c_{i2}+\text{H.c.}\right)+V\rho_{i1}\rho_{i2}\Bigg], \end{aligned} \tag{14}$$

where $c_{i\alpha}^{\dagger}$ creates a spinless electron at site i of chain α, $\rho_{i\alpha} \equiv c_{i\alpha}^{\dagger}c_{i\alpha}$, and $\varepsilon_{i\alpha}$ is a random number in the interval $[-\varepsilon/2, \varepsilon/2]$. The interchain hopping determines the subband splitting Δ between the symmetric and antisymmetric states. The rms amplitude of the intersubband scattering is $\varepsilon/2\sqrt{3}$, and the expectation value of the difference in subband populations is given by

$$\langle K_1 - K_2\rangle = -2\partial E_0/\partial\Delta. \tag{15}$$

Figure 1 shows the average subband occupancies and persistent current for an ensemble of 100 rings with 5 spinless electrons on 18 sites as a function of V, calculated using the Lanczos technique. The subband splitting $\Delta = 0.8$ is chosen so that in the absence of disorder and interactions, $K_1 = 3$ and $K_2 = 2$, leading to a large cancellation of the persistent current due to the different parities of the two channels. The on-site disorder $\varepsilon = 2 > \Delta, E_F$ mixes the two channels, but does not lead to strong backscattering (localization). Figure 1(a) indicates that the average difference in subband occupancies increases from 1.05 at $V = 0$ to 4.55 and $V = 20$, corresponding to a 96% condensation into the lowest subband. As the intersubband polarization increases, the first harmonic of the persistent current oscillates in sign and increases in magnitude, similar to the behavior of the integrable system given in Eq. (13). In contrast, the second harmonic remains paramagnetic, and does not exhibit a noticeable increase. The statistical width of the current distribution is $\delta I = (\int_0^{2\pi}[\langle I(\phi)^2\rangle - \langle I(\phi)\rangle^2]d\phi/2\pi)^{1/2} = (0.125 \pm .01)I_0$ for the whole range of V. Note that $|\langle I_1\rangle| \ll \delta I$ for $V = 0$, while $\langle I_1\rangle$ is diamagnetic with $|\langle I_1\rangle| \gg \delta I$ for $V > 5$, as expected for a system with N odd due to parity locking.

Figure 1 shows that the parity-locking effect is not significantly modified even in systems where the intersubband scattering is comparable to the subband splitting. While the subband occupancies are no longer constants of the motion in a disordered system, there is a corresponding topological invariant, namely, the number of transverse nodes in the many-body wave function [2] (i.e., nodes

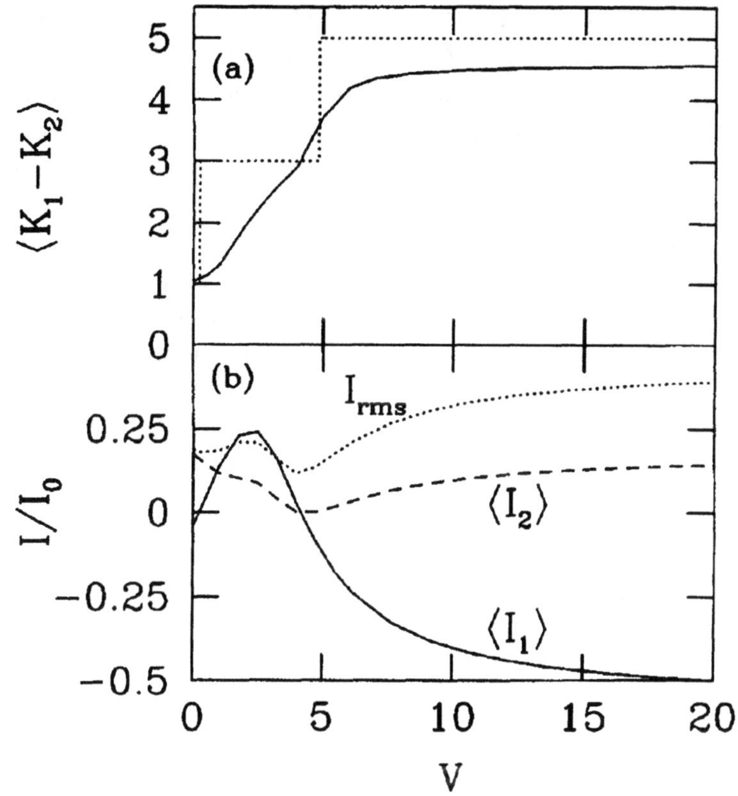

Figure 1. Results for an ensemble of 100 disordered two-channel rings with 5 spinless electrons on 18 sites as a function of the interchain interaction V. The subband splitting is $\Delta = 0.8$. (a) Difference of average subband occupancies $\langle K_1 - K_2 \rangle = -2\partial E_0/\partial\Delta$ at $\phi = \phi_0/4$ for disorder parameter $\varepsilon = 2$ (solid curve) and $\varepsilon = 0$ (dotted curve). (b) Average first and second harmonics $\langle I_1 \rangle$ and $\langle I_2 \rangle$ of the persistent current and $I_{\rm rms} = (\int_0^{2\pi} \langle I(\phi)^2 \rangle d\phi/2\pi)^{1/2}$ for $\varepsilon = 2$. The relation of $\langle I_1 \rangle$ to $\langle K_1 - K_2 \rangle$ is qualitatively similar to that in the model with unbroken $SU(2)$ symmetry, Eq. (13).

which encircle the AB flux ϕ). The lowest subband has no such nodes, while each electron in the second subband contributes one transverse node. In order for two electrons to pass each other as they circle the ring, such a transverse node must be present. As V increases, it becomes energetically unfavorable for electrons to approach each other, so transverse nodes in the many-body wave function will tend to be suppressed. In the strongly-correlated limit, all such nodes will be eliminated, leading to a state whose parity depends only on the *total* number of electrons. In such a state, the persistent currents of all channels add constructively, leading to a large enhancement of I_1.

In conclusion, we have proposed a novel interaction effect which leads to a large enhancement of the persistent current (and in particular, of its first harmonic) in multichannel rings due to correlations in the contributions of different channels. Sufficiently strong interactions were shown to lead to an enhancement of the typical persistent current by a factor of $M^{1/2}$ in a ballistic ring with M channels, compared to the value for noninteracting electrons. It was shown that even when interactions are weak compared to those necessary to enforce complete parity-locking, as is likely to be the case in metallic rings such as those studied in Refs. [6] and [7], the persistent current may still be substantially enhanced by interchannel correlations. It should be emphasized that the parity-locking effect is a pure interaction effect in multichannel rings; disorder plays no essential role. It is therefore complementary to mechanisms previously proposed for the enhancement of the persistent current [11, 12, 13, 14, 15], which rely on the competition between disorder and interactions. It is likely that both such mechanisms are important to explain the anomalously large observed value [6, 7] of the persistent current in disordered metallic rings.

Acknowledgments

C. A. S. thanks M. Büttiker and F. Hekking for valuable discussions. This work was supported by the Swiss National Science Foundation.

References

[1] N. Byers and C. N. Yang, Phys. Rev. Lett. **7**, 46 (1961).

[2] A. J. Leggett, in *Granular Nanoelectronics*, edited by D. K. Ferry *et al.* (Plenum, New York, 1991), p. 297.

[3] D. Loss, Phys. Rev. Lett. **69**, 343 (1992).

[4] M. Buttiker, Y. Imry and R. Landauer, Phys. Lett. **96A**, 365 (1983).

[5] H. F. Cheung, E. K. Riedel, and Y. Gefen, Phys. Rev. Lett. **62**, 587 (1989).

[6] L. P. Lévy, G. Dolan, J. Dunsmuir and H. Bouchiat, Phys. Rev. Lett. **64**, 2074 (1990).

[7] V. Chandrasekhar *et al.*, Phys. Rev. Lett. **67**, 3578 (1991).

[8] D. Mailly, C. Chapelier and A. Benoit, Phys. Rev. Lett. **70**, 2020 (1993).

[9] B. Reulet, M. Ramin, H. Bouchiat, and D. Mailly, Phys. Rev. Lett. **75**, 124 (1995).

[10] H. Bouchiat and G. Montambaux, J. Phys. (France) **50**, 2695 (1989); F. von Oppen and E. K. Riedel, Phys. Rev. Lett. **66**, 84 (1991); B. L. Altshuler, Y. Gefen, and Y. Imry, *ibid.* **66**, 88 (1991).

[11] U. Eckern, Z. Phys. B **82**, 393 (1991).

[12] A. Schmid, Phys. Rev. Lett. **66**, 80 (1991).

[13] D. Yoshioka and H. Kato, Physica B **212**, 251 (1995); G. Bouzerar and D. Poilblanc, J. Phys. (France) (to be published).

[14] T. Giamarchi and B. S. Shastry, Phys. Rev. B **51**, 10 915 (1995); H. Mori and M. Hamada, *ibid.* **53**, 4850 (1996).

[15] R. Berkovits and Y. Avishai, Europhys. Lett. **29**, 475 (1995); G. Bouzerar and D. Poilblanc, Phys. Rev. B **52**, 10 772 (1995); G. Chiappe, E. Louis, and J. A. Vergés, Solid State Commun. **99**, 717 (1996).

[16] B. Sutherland, Phys. Rev. A **4**, 2019 (1971); **5**, 1372 (1972).

[17] C. A. Stafford and D. F. Wang (unpublished).

[18] B. Sutherland, Phys. Rev. Lett. **20**, 98 (1968).

[19] H. Frahm and V. E. Korepin, Phys. Rev. B **43**, 5653 (1991); J. Carmelo, P. Horsch, P. A. Bares, and A. A. Ovchinnikov, *ibid.* **44**, 9967 (1991).

[20] F. V. Kusmartsev, J. Phys.: Condens. Matter **3**, 3199 (1991).

[21] N. Yu and M. Fowler, Phys. Rev. B **45**, 11 795 (1992).

[22] H. Shiba, Phys. Rev. B **6**, 930 (1972).

[23] Additional *intrachain* interactions were not found to modify the parity-locking effect in an essential way, but do lead to an enhancement of localization in models with spinless electrons.

GENERALIZING MERTON'S APPROACH OF PRICING RISKY DEBT: SOME CLOSED-FORM RESULTS

D. F. Wang

Department of Statistics and Actuarial Science
University of Waterloo
Waterloo, Ont., N2L 3E5
Canada

Toronto Dominion Bank
Canada
d6wang@barrow.uwaterloo.ca

Abstract In this work, I generalize Merton's approach of pricing risky debt to the case where the interest rate risk is modeled by the CIR term structure. Closed-form result for pricing the debt is given for the case where the firm value has non-zero correlation with the interest rate. This extends previous closed-form pricing formulae of zero-correlation case to the generic one of non-zero correlation between the firm value and the interest rate.

PACS: 05.40.+j; 01.90+g

Keywords: Defaultable bond; CIR term structure

One well-known approach of pricing risky debt was pioneered by Merton a long time ago. Within the framework, one assumes a stochastic process for the firm value and treats the risky debt as a option [1, 2, 3, 4, 5, 6]. In Merton's original work [1], he assumed that the interest rate is constant and that the default event of the debt can only occur at the time of maturity. Closed-form results for pricing the risky debt were given explicitly. As he has pointed out, one can also study the case when stochastic interest rate is taken into account. This can easily be achieved by using Merton's work that generalized Black-Scholes formula of the option pricing with stochastic interest rate [1].

However, one has to make the assumption that the bond process of the stochastic interest rate has a non-stochastic volatility which is allowed to be deterministic time dependent [1]. Therefore, Shimko et al. [2] applied Merton's results to the case of stochastic interest rate described by the Vasicek model, the bond process of which has a non-stochastic volatility. However, they were unable to handle the case where the interest rate is modeled with the CIR term structure, as the CIR interest rate model will give rise to a bond process having stochastic volatility [7]. It remained open whether one can give a similar closed form results for the risky debt when the interest rate risk is modeled with CIR term structure. In one recent work, I gave the closed-form pricing formulae for the risky debt in the case where CIR term structure is used for the default-free interest rate. However, using the moment generating functional, one has to assume that the firm value is uncorrelated with the interest rate [8]. In this paper, I extend my closed-form results of pricing risky debt to the generic case where the correlation between the firm value and the interest rate is non-zero.

Let us first assume a probability space denoted by $(\Omega, P, \{F_t\}, F)$, with the filtration $\{F_t\}$. Consider the value of the firm that is described by the following process:

$$\frac{dV}{V} = \mu\, dt + \sigma\, dZ_1, \tag{1}$$

where Z_1 is a Brownian motion in the probability space. The interest rate process is assumed to be the one given by Cox-Ingersoll-Ross [7]:

$$dr = (a - \beta r) + \eta dZ_2, \tag{2}$$

where $\eta = \sigma_r r^{1/2}$ with σ_r as a constant. The coquadratic variational process is $[Z_1, Z_2] = \rho t$. The correlation coefficient ρ is a non-zero constant during the following consideration.

The assumptions in Merton's paper are also made here [1]. The firm value is assumed to be independent of its capital structure by assuming that MM theorem is valid. The firm issues debt and equity. The total value of the firm is the sum of equity and debt. The PDE satisfied by the equity is given by

$$H_\tau = \frac{\sigma^2}{2}V^2 H_{VV} + \rho\eta\sigma V H_{Vr} + \frac{\eta^2}{2}H_{rr} + rVH_V + (\alpha - \beta r)H_r - rH, \tag{3}$$

where $H = H(V, r, T-t)$ and $\tau = T-t$ is time to the maturity, and α is sum of a plus the constant representing the market price of the interest rate risk. At $\tau = 0$, the equity should satisfy the boundary condition that $H = \max(0, V(T) - B)$, where B is the face value of the debt issued by the firm maturing at time T. The risky debt price is therefore given by $Y = V(t) - H(V, r, T-t)$. For simplicity, it is assumed here, as Merton did, that event of default of the risky debt can only occur at the time of maturity.

Following the standard risk neutral approach, we write the equity price as:

$$H = E^Q(e^{-\int_t^T r(s)ds}\max(0, V(T) - B)|F_t), \tag{4}$$

where the expectation E^Q means that in the risk-neutral-adjusted world. In this risk-neutral world, the firm value and the interest rate will follow the stochastic differential equations as

$$\begin{aligned} d\ln V &= (r - \tfrac{1}{2}\sigma^2)\,dt + \sigma\,d\hat{Z}_1 \\ dr &= (\alpha - \beta r)\,dt + \eta\,d\hat{Z}_2. \end{aligned} \tag{5}$$

Here, both $\hat{Z}_1$ and $\hat{Z}_2$ are Wiener processes in the risk-neutral world, and the co-quadratic process is $[\hat{Z}_1, \hat{Z}_2] = \rho t$. In principle, one can go on with Monte-Carlo simulation based on Eqs. (4) and (5). However, we are interested in finding closed-form result for the equity price here.

In the risk-neutral world, the two Brownian motions $\hat{Z}_1$ and $\hat{Z}_2$ can be represented in the following way. Suppose that X and Y are two independent Brownian motions in the risk-neutral world. We can find such independent Brownian motions that $\hat{Z}_1 = \rho X + \sqrt{(1-\rho^2)}Y$, and $\hat{Z}_2 = X$. The stochastic differential equation for the firm value is governed by

$$d\ln V = [r\,dt + \sigma\rho\,dX - \tfrac{1}{2}\sigma^2\rho^2\,dt] - \tfrac{1}{2}\sigma^2(1-\rho^2)\,dt + \sqrt{(1-\rho^2)}\sigma\,dY, \tag{6}$$

where one is working in the risk-neutral world.

Conditional on that the sample path $\{X\}$ of the Brownian motion X is given for time interval $[t, T]$, let us consider the following expectation:

$$\begin{aligned} h &= E^Q(e^{-\int_t^T r(s)\,ds}\max(V(T) - B, 0)|F_t, \{X\}) \\ &= e^{-\int_t^T r(s)\,ds} E^Q(\max(V(T) - B, 0)|F_t, \{X\}). \end{aligned} \tag{7}$$

This conditional expectation can be computed by standard way as dealing with Black-Scholes case. It is found that

$$\begin{aligned} h = {} & e^{\sigma\rho\int_t^T dX} e^{-(1/2)\sigma^2\rho^2(T-t)} \\ & \times[V(t)\mathrm{N}(d_1) - Be^{-(\int_t^T r(s)\,ds - (1/2)\sigma^2\rho^2(T-t) + \sigma\rho\int_t^T dX)}\mathrm{N}(d_2)], \end{aligned} \tag{8}$$

where $\mathrm{N}(d)$ is the standard accumulative function of normal distribution $\mathrm{N}(0,1)$ and

$$d_{1,2} = \frac{\ln(V(t)/B) + [\int_t^T r(s)ds - \tfrac{1}{2}\rho^2\sigma^2(T-t) + \sigma\rho\int_t^T dX \pm \tfrac{1}{2}\sigma^2(1-\rho^2)(T-t)]}{\sigma\sqrt{(1-\rho^2)(T-t)}} \tag{9}$$

Let us denote $g(m, W)$ the probability density for $\int_t^T r(s)\,ds$ to be in the region $[m, m + dm]$, conditional on that $W = \int_t^T dX$ is fixed. In the following we will show that $g(m, W)$ is independent of the variable W. The equity price of the firm can be represented as

$$H = \int_{-\infty}^{+\infty} P(W)\, dW \int_{0}^{+\infty} g(m)\, dm \times [e^{\sigma\rho W} e^{-(1/2)\sigma^2\rho^2(T-t)} V(t) \mathrm{N}(d_1) - Be^{-m}\mathrm{N}(d_2)], \quad (10)$$

where $P(W)$ is normal distribution density, $P(W) = \exp(-0.5W^2/(T - t))/\sqrt{2\pi(T-t)}$, and

$$d_{1,2} = \frac{\ln(V(t)/B) + [m - \frac{1}{2}\rho^2\sigma^2(T-t) + \sigma\rho W \pm (1/2)\sigma^2(1-\rho^2)(T-t)]}{\sigma\sqrt{(1-\rho^2)(T-t)}} \quad (11)$$

The distribution function $g(m)$ can be found from the moment generating functional. For the Cox-Ingersoll-Ross term structure, the density function $g(m)$ can be found easily. Consider the moment-generating function

$$I(x) = E^Q(e^{-x\int_t^T r(s)\, ds} | F_t) = \int_0^\infty e^{-mx} g(m)\, dm, \quad (12)$$

where x is any non-zero numbers. Using the bond price of Cox-Ingersoll-Ross, we find the moment generating function. For CIR term structure, stochastic differential equation governing the short rate is $dr = (\alpha - \beta r)\, dt + \sigma_r \sqrt{r}\, d\hat{Z}_2$, will remain unchanged under the scaling transformation:

$$r \to xr, \qquad \beta \to \beta, \qquad \alpha \to x\alpha, \qquad \sigma_r \to x^{1/2}\sigma_r, \quad (13)$$

where x is any positive real number. Denote $D(r(t); \alpha, \beta, \sigma_r, T - t) = E^Q(e^{-\int_t^T r(s)\, ds} | F_t)$ the riskless zero-coupon bond price at time t whose payoff at maturity T is one. The exact closed form of this zero-coupon bond price was provided explicitly [7]. We obtain

$$I(x) = D(xr(t), x\alpha, \beta, x^{1/2}\sigma_r, T - t) = \int_0^\infty g(m) e^{-xm} dm. \quad (14)$$

Taking the inverse transformation, we will be able to find the density function $g(m)$.

In order to see why $g(m, W)$ is independent of W, we first look at the discrete version of $\int_t^T r(s)\, ds$. In the discrete version, we see that W dependence will only make higher-order contribution. Therefore, in the continuous limit, the density distribution of $\int_t^T r(s)\, ds$ for given W will not have W dependence. This is why we can represent the equity price in terms of two double integral as Eq. (10). The price of the risky debt that pays B dollars at time of maturity T is simple given by $V(t) - H$.

In summary, we have generalized Merton's approach of pricing risky debt to the situation where the interest rate risk is modeled with the CIR term structure. Exact closed forms for pricing risky debt are provided explicitly. This goes beyond the situation where the riskless bond process has non-stochastic volatility (such as for Vasicek interest rate model) and the option pricing can be handled by Merton's generalized Black-Scholes method.

Acknowledgments

I am indebted to Professors P. Boyle and D. McLeish for the finance theories I learned from them. Conversations with Prof. K.S. Tan of UW, Dr. Hou-Ben Huang and Dr. Z. Jiang of TD Securities, Dr. Bart Sisk and Dr. A. Benn of TD Bank, Dr. Daiwai Li and Dr. Craig Liu of Royal Bank, Dr. ChongHui Liu and Dr. J. Faridani of Scotia Bank, are gratefully knowledged. I also wish to thank Dr. Rama Cont for informative communication. The opinions of this article are those of the author's, and they do not necessarily reflect the institutions the author is affiliated with. Any errors of this article are mine.

References

[1] R. C. Merton, On the pricing of corporate debt: the risk structure of interest rates, J. Finance **29**, 449 (1974).

[2] D. Shimko, N. Tejima and D.R. Van Deventer, The pricing of risk debt when interest rates are stochastic, J. Fixed Income, 58, Sept. (1993).

[3] F. Black and J.C. Cox, Valuing corporate securities: some effects of bond indenture provisions, J. Finance **31**, 351 (1976).

[4] F. Longstaff and E. Schwartz, A simple approach to valuing risky fixed and floating rate debt, J. Finance **50**, 789 (1995).

[5] C. Zhou, preprint of Board of Governers of Federal Reserve, 1998.

[6] O. Vasicek, An equilibrium characterization of the term structure, J. Financial Econom. **5**, 177 (1977).

[7] J.C. Cox, J.E. Ingersoll and S.A. Ross, A theory of the term structure of interest rates, Econometrica **53**, 385 (1985).

[8] D.F. Wang, preprint of TD Bank and Uiv. of Waterloo, May 1998, to be published.

FSC
www.fsc.org
MIX
Papier aus verantwortungsvollen Quellen
Paper from responsible sources
FSC® C105338

If you have any concerns about our products,
you can contact us on
ProductSafety@springernature.com

In case Publisher is established outside the EU,
the EU authorized representative is:
Springer Nature Customer Service Center GmbH
Europaplatz 3, 69115 Heidelberg, Germany

Printed by Libri Plureos GmbH
in Hamburg, Germany